Springer

Tokyo
Berlin
Heidelberg
New York
Barcelona
Hong Kong
London
Milan
Paris
Singapore

M.D. Cooper, T. Takai, J.V. Ravetch (Eds.)

Activating and Inhibitory Immunoglobulin-like Receptors

With 52 Figures, Including 13 in Color

Springer

Max D. Cooper, M.D.
Howard Hughes Medical Institute
University of Alabama at Birmingham
378 Wallace Tumor Institute
UAB Station, Birmingham, Alabama 35294, USA

Toshiyuki Takai, Ph.D.
Department of Experimental Immunology
Institute of Development, Aging and Cancer
Tohoku University
Seiryo 4-1, Sendai 980-8575, Japan

Jeffrey V. Ravetch, M.D., Ph.D.
Laboratory of Molecular Genetics and Immunology
The Rockefeller University
New York 10021-6399, USA

The image on the front cover shows degranulating mast cells and is the symbol of the CREST International Symposium on Immunoglobulin-like Receptors held in Sendai, Japan.

The publication of this book is supported by The CREST Program of Japan Science and Technology Corporation

ISBN 4-431-70297-0 Springer-Verlag Tokyo Berlin Heidelberg New York

Library of Congress Cataloging-in-Publication Data

Activating and inhibitory immunoglobulin-like receptors / [edited by] M.D. Cooper, T. Takai, J.V. Ravetch.
     p. ; cm.
  Includes bibliographical references and index.
  ISBN 4431702970 (hard cover : alk. paper)
    1. Fc receptors. 2. Immunoglobulins. 3. Cell receptors. 4. Immune response--Regulation. I. Cooper, Max D., 1933- II. Takai, T. (Toshiyuki), 1958- III. Ravetch, J.V. (Jeffrey V.), 1951-
  [DNLM: 1. Receptors, Fc--immunology. 2. Immune System--physiology. 3. Immunoglobulins--physiology. 4. Receptors, Fc--physiology. QW 601 A188 2001]
  QR185.8.F33 A26 2001
  616.07'9--dc21

                                                            2001020221

Printed on acid-free paper

Typesetting: Camera-ready by the editors and authors
Printing and binding: Hirakawa-kogyo, Japan
SPIN: 10760872

Preface

A remarkable spectrum of novel immunoreceptors sharing related immunoglobulin-like domains and signaling potential has been identified in recent years. These immunoglobulin superfamily members include the killer immunoglobulin-like receptors (KIR) expressed by the NK cells and NK-T cells in humans; the immunoglobulin-like transcripts (ILT)/leukocyte immunoglobulin-like receptors (LIR) expressed by T, B, NK, dendritic, and myeloid cells in humans; and the paired immunoglobulin-like receptors found on B lymphocytes, dendritic cells, and myeloid cells in rodents. These novel receptors have attracted widespread interest because they resemble the TCR, BCR, and FcR complexes in their ability to serve as activating or inhibitory receptors on the cells that bear them. Moreover, they are well positioned to affect both innate and adaptive immunity. Still, we do not know the full spectrum of ligands for these new receptor families, and our understanding of their physiological roles is far from complete. At this juncture in a dynamic field of research, the CREST International Symposium on Immunoglobulin-like Receptors was held September 18–19, 2000, at the Sendai International Center in Sendai City, Japan. The aim was to bring together leading scientists in the field to discuss currently available information on the newest immunoglobulin-like receptors and what is known about the TCR, BCR, FcR, and lectin-type receptors. The integrated picture emerging during this symposium predicts that genetic or therapeutic modifications of these immunoglobulin-like receptors may alter the development or clinical course of a variety of disease states, including hypersensitivity disorders, autoimmune diseases, and cancer.

This book provides a topical overview of the roles and characteristic features of the immunoglobulin-like receptors and related molecules in the immune system. The editors wish to thank each of the contributors for their participation in the CREST symposium and for their excellent review articles.

January 2001

Max D. Cooper, Toshiyuki Takai, and Jeffrey V. Ravetch

Contents

1 *Introduction*
Fc receptors: Activation-inhibition receptor pairing

Jeffrey V. Ravetch

Laboratory of Molecular Genetics and Immunology, The Rockefeller University, New York 10021-6399,USA

Cell surface receptors for IgG immune complexes are found on essentially all major subgroups of cells of the immune system, including antigen presenting cells, such as dendritic cells, antibody producing cells, such as B lymphocytes, and effector cells such as macrophages, granulocytes, mast cells and NK cells (Ravetch and Kinet 1991; Daëron 1997; Ravetch and Clynes 1998). Crosslinking of these receptors triggers a wide array of cellular responses which, in concert, contribute significantly to shaping the antibody repertoire, maintaining tolerance and initiating the inflammatory response. The ability to regulate this wide array of biological responses is central to the generation of appropriate immunity in response to challenge by infectious pathogens, senescent cells and neoplasms. The importance of maintaining appropriate responses to immune complexes is revealed by mutations in these pathways which result in the emergence of systemic autoimmunity and autoimmune diseases. The ability to regulate these potent biological systems is achieved by coupling receptors with similar ligand specificity to opposing cellular responses. Thus, immune complexes co-engage both activation (FcγRIII) and inhibitory (FcγRIIB) receptors simultaneously, resulting in a cellular response which reflects the expression levels of each member of the pair (Bolland and Ravetch 1999). In this manner, FcRs are capable of setting thresholds for activation by immune complexes, terminating activation signals and modulating B cell responses during antigen driven affinity maturation.

The inhibitory FcR, FcγRIIB, terminates activation responses upon its co-ligation to an ITAM-containing receptor, such as the BCR or FcR. Phosphorylation of the ITIM-motif in RIIB results in the recruitment of the inositol polyphosphate phosphatase SHIP, the hydrolysis of PIP3 and the release of PH-domain containing proteins Btk and Plcγ, culminating in the specific inhibition of calcium influx into the activated cell (Bolland et al. 1998). In addition to this inhibitory pathway, clustering of RIIB on B lymphocytes results in the triggering of apoptosis (Pearse et al. 1999). This dual activity for RIIB leads to an hypothesis to account for the selection of somatically hypermutated B

1

lymphocytes during affinity maturation in the germinal center. The interaction of these B lymphocytes with immune complexes retained on follicular dendritic cells can lead to either negative selection by RIIB engagement in the absence of BCR co-engagement or survival and selection, depending upon the strength of the BCR interaction with antigen. The significance of this dual function for RIIB on B cells is revealed by gene disruption studies. C57Bl/6 mice deficient for RIIB display a severe autoimmune phenotype, with anti-dsDNA and chromatin antibodies appearing by 4 months of age (Bolland and Ravetch 2000). Deposition of immune complexes results in glomerulonephritis with the animals succumbing to this autoimmune disease by 8 months of age. The disease is B cell autonomous and can be transferred to B cell deficient mice with B cells from C57Bl/6 RIIB deficient B cells. In contrast, Balb/C mice deficient in RIIB have normal lifespans and are not susceptible to spontaneous autoimmunity. This strain-specific phenotype for RIIB deficiency indicates that this gene behaves as a susceptibility factor under the influence of epistatic modifiers. As such, it displays many of the characteristics of genes likely to contribute to systemic autoimmunity in outbred populations and is a candidate gene for susceptibility to systemic autoimmune diseases like lupus.

RIIB also modulates B cell responses in induced autoimmune disease models, like collagen induced arthritis (CIA) and Goodpasture s syndrome (see the chapter by T. Takai, this volume). Immunization of RIIB deficient mice with type II collagen overrides the MHC dependence seen in CIA by triggering the appearance of murine anti-collagen autoantibodies (Yuasa et al. 1999). Similarly, immunization of RIIB deficient mice with type IV collagen results in hemorrhagic pulmonary disease and autoimmune glomerulonephritis with the characteristic basement membrane deposition pattern seen only Goodpasture s (Nakamura et al. 2000). Here, too, autoreactive anti-type IV collagen antibodies appear only in RIIB deficient mice, indicating the importance of this receptor in maintaining peripheral B cell tolerance.

RIIB is critical for determining the level of activation of myeloid cells by cytotoxic antibodies and immune complexes. Cytotoxic antibodies to tumor antigens are capable of mediating $Fc\gamma RIII$ dependent ADCC in vivo in both syngenic and xenograft models of antibody-mediated tumor arrest (Clynes et al. 1998; Clynes et al. 2000). The absolute level of cytotoxicity is determined by RIIB expression. Disruption of RIIB expression or signaling enhances the potency of these antibodies by 10-100 fold. In a similar manner, immune complexes trigger inflammatory responses in models of cutaneous, alveolar and renal Arthus reactions by the co-ordinate engagement of both $Fc\gamma RIII$ and $Fc\gamma RIIB$ (Clynes et al. 1999). Down modulation of RIII expression or function uncouples immune complexes from activation of inflammatory responses in vivo, thus protecting animals from the potentially pathogenic effects of these molecules in systemic autoimmunity (Clynes et al. 1998). Conversely, reducing the function of the inhibitory receptor, RIIB, by gene disruption or blocking antibodies enhances the severity of inflammatory responses and lowers the threshold required for triggering effector responses.

This systematic dissection of the opposing functions of RIIB and RIII in vivo offers a paradigm for pathways used by the immune system for maintaining homeostatic control over potent cellular responses. This paradigm, demonstrated for Fc receptors, will likely be true for the growing family of ITAM-ITIM receptor pairs. By coupling similar ligand binding activities with opposing signaling molecules and regulating the coordinate expression of these pairs of receptors, a threshold can be raised or lowered along with the absolute magnitude of the elicited response. By regulating the ratio of activation to inhibitory receptor expression, a specific cell can either be responsive or inert in response to ligand stimulation. The factors which regulate this balance for FcRs contribute to the maintenance of tolerance and the activation of effector cells and are thus likely to be important modifiers for autoimmune diseases.

References

Bolland S, Pearse R, Kurosaki T, Ravetch JV (1998) SHIP modulates immune receptor responses by regulating membrane association of Btk. Immunity 8:509-516

Bolland S, Ravetch JV (1999) Inhibitory pathways triggered by ITIM-containing receptors. Adv Immunol 72:149-177

Bolland S, Ravetch JV (2000) Spontaneous autoimmune disease in FcγRIIB-deficient mice results from strain-specific epistasis. Immunity 13:277-285

Clynes RA, Dumitru C, Ravetch JV (1998) Uncoupling of immune complex formation and kidney damage in autoimmune glomerulonephritis. Science 279:1052-1054

Clynes R, Maizes JS, Guinamard R, Ono M, Takai T, Ravetch JV (1999) Modulation of immune complex induced inflammation in vivo by the coordinate expression of activation and inhibitory Fc receptors. J Exp Med 189:179-186

Clynes R, Ravetch JV (1995) Cytotoxic antibodies trigger inflammation through Fc receptors. Immunity 3:21-26

Clynes R, Takechi Y, Moroi Y, Houghton A, Ravetch JV (1998) Fc receptors are required in passive and active immunity to melanoma. Proc Natl Acad Sci USA 95:652-656

Clynes RA, Towers T, Presta LG, Ravetch JV (2000) Inhibitory Fc receptors modulate in vivo cytoxicity against tumor targets. Nature Med 6:443-446

Daëron M (1997) Fc receptor biology. Annu Rev Immunol 15:203-234

Nakamura A, Yuasa T, Ujike A, Ono M, Nukiwa T, Ravetch JV, Takai T (2000) Fcγ receptor IIB-deficient mice develop Goodpasture's syndrome upon immunization with type IV collagen: A novel murine model for autoimmune glomerular basement membrane disease. J Exp Med 191:899-906

Pearse RN, Kawabe T, Bolland S, Guinamard R, Kurosaki T, Ravetch JV (1999) SHIP recruitment attenuates FcγRIIB-induced B cell apoptosis. Immunity 10:753-760

Ravetch JV, Clynes RA (1998) Divergent roles for Fc receptors and complement in vivo. Annu Rev Immunol 16:421-432

Ravetch JV, Kinet J-P (1991) Fc receptors. Annu Rev Immunol 9:457-492

Yuasa T, Kubo S, Yoshino T, Ujike A, Matsumura K, Ono M, Ravetch JV, Takai T (1999) Deletion of FcγRIIB renders H-2^b mice susceptible to collagen-induced arthritis. J Exp Med 189:187-194

Part A

Gene Structure, Expression, and Evolution

2
Phylogeny of Paired Ig-like Receptors

Glynn Dennis Jr., Ching-Cheng Chen, Hiromi Kubagawa, and Max D. Cooper

Division of Developmental and Clinical Immunology, Departments of Microbiology, Pathology, Pediatrics and Medicine, and the Howard Hughes Medical Institute, University of Alabama at Birmingham, AL 35294-3300, USA

Summary. The mouse paired immunoglobulin-like receptors of activating (PIR-A) and inhibitory (PIR-B) types exemplify an emerging family of hematopoietic cell surface receptors whose roles in natural and acquired immunity have yet to be fully elucidated. As implied by their name, coordinate expression of PIR-A and PIR-B is seen on myeloid, dendritic, and B cell lineages. PIR-A and PIR-B share similar ligand-binding extracellular regions composed of six Ig-like domains, but they differ in their signaling potentials on the basis of differences in their transmembrane and cytoplasmic regions. Although PIR-A and PIR-B themselves are not conserved beyond rodents, PIR homologs have been identified in other mammals and birds, suggesting that the PIR genes belong to an ancient ancestral lineage of related Ig-like receptor families. A phylogenetic analysis, employing primary amino acid sequences and three-dimensional protein structures of the PIR homologs and their relatives, suggests a common ancestry for the PIR and Fc receptor families.

Key words. Ig-like receptors, Phylogeny, Leukocyte receptor complex, Innate immunity, Fc receptor

Mouse PIR-A and PIR-B

The fortuitous discovery of mouse PIR-A and PIR-B resulted from the search for mouse homologs of the human IgA Fc receptor (FcαR) (Hayami et al. 1997; Kubagawa et al. 1997). The FcαR has two Ig-like domains in its extracellular (EC) region and a short cytoplasmic tail. A charged arginine residue in the transmembrane region promotes its association with a signal transducing molecule, the Fc receptor common γ chain (FcRγc) that is shared by other Fc receptors, including FcϵRI, FcγRI, and FcγRIII, as well as the T cell receptor (TCR)/CD3 complex expressed by minor subpopulations of T cells (Morton et al.

1996). An immunoreceptor tyrosine-based activation motif (ITAM) in the FcRγc transmembrane adaptor chain facilitates cellular activation via these complex cell surface receptors. Like human FcαR, mouse PIR-A has a potential ligand-binding region composed of Ig-like domains, and a transmembrane region that contains a positively charged arginine residue. Similarly, the cytoplasmic regions of PIR-A and FcαR are relatively short and lack recognizable signaling motifs. Through its positively charged transmembrane region PIR-A associates with the FcRγc to form a receptor complex capable of promoting cellular activation through its interaction with intracellular protein tyrosine kinases (Kubagawa et al. 1999; Maeda et al. 1998a; Ono et al. 1999; Yamashita et al. 1998b). PIR-A differs from the FcαR in that it possesses a relatively large EC region with six Ig-like domains. Moreover, PIR-A isoforms that differ in their EC region sequences are encoded by approximately eight different genes.

PIR-B, the inhibitory partner of PIR-A, shares greater than 90% amino acid identity in its EC region with PIR-A, but has an uncharged transmembrane segment and a relatively long cytoplasmic tail. Of the five tyrosine residues present in the cytoplasmic region of PIR-B (Y713, Y742, Y770, Y794 and Y824), three (Y713, Y794 and Y824) are embedded in amino acid sequences that conform to the consensus sequence of immunoreceptor tyrosine-based inhibitory motifs (ITIM). Mutational analysis indicates that simultaneous mutation of both Y794 and Y824 significantly impairs the inhibitory function of PIR-B (Bléry et al. 1998; Yamashita et al. 1998b). PIR-B ITIMs are constitutively tyrosine-phosphorylated in vivo and serve as docking sites for protein tyrosine phosphatases like the src-homology phosphatase-1 (SHP-1) or SHP-2, which in turn can inhibit cellular activation signals generated through protein tyrosine kinase pathways (Ho et al. 1999; Maeda et al. 1998b; Maeda et al. 1999). A third functional ITIM-like motif (Y742) has recently been identified, which may activate a different signaling pathway (our unpublished results).

PIR-A and PIR-B are coordinately expressed on B lymphocytes, dendritic cells, monocyte/macrophages, granulocytes, and platelets, but have not yet been found on T cells and NK cells. Cell surface expression of PIR molecules on B lymphocytes and myeloid cells is up-regulated during cellular maturation. Examination of the different splenic subpopulations of B cells indicates that the surface PIR molecules expression levels are highest on marginal zone B cells. Moreover, the B1 B cells express higher PIR levels than the B2 B cells. The implication that activated B cells express higher PIR levels is supported by the observation that LPS stimulation up-regulates PIR expression levels on B cells and myeloid cells (Kubagawa et al. 1999). Interestingly, although mast cells also produce both PIR-A and PIR-B, PIR-B molecules are predominantly expressed on the mast cell surface. This finding implies that in mast cells there is limited availability of the FcRγc chains, which are required for PIR-A expression on the cell surface (Kubagawa et al. 1999).

Analysis of the *Pirb* gene sequence and of PIR-A and PIR-B cDNA sequences in different mouse strains indicates the polymorphic nature of the *Pira* and *Pirb* genes. The approximately eight *Pira* genes also encode slightly different PIR-A isoforms. Isoform EC domain differences include extended sequence motifs within the Ig-like domains that resemble the complementary determining regions of antibodies (Chen et al. 1999; Kubagawa et al. 1997). Although PIR ligands are currently undefined, this type of variability suggests the possibility of PIR-A and PIR-B gene coevolution with their polymorphic ligands.

Rat PIR-A and PIR-B homologs

The overall structure of the rat (ra) PIR-A and raPIR-B molecules is remarkably similar to that of their mouse homologs. The EC region of raPIR-A is composed of six Ig-like domains. The raPIR-A molecules also have a transmembrane region endowed with a positively charged arginine residue and a short cytoplasmic tail with no recognizable signaling motifs. The raPIR-B molecules shares greater than 90% identity in its EC region with raPIR-A, but has an uncharged transmembrane region and a relatively long cytoplasmic tail that has four candidate ITIMs. Southern blot analysis of rat genomic DNA suggests that a single *raPirb* gene shares similarity in its EC region with multiple *raPira* genes. The high degree of similarity between the rat and mouse *Pir* genes suggests an orthologous relationship. However, the paired expression rule of mouse PIR-A and PIR-B is broken in rats, wherein unpaired expression of raPIR-A may occur in B cells and raPIR-B alone has been observed in NK cells (Dennis et al. 1999). These findings suggest that, while very similar, the elements regulating expression of rat and mouse PIR genes have diverged since their last common ancestor some 20 million years ago.

PIR homologs in humans

The mouse *Pir* genes reside near the proximal end of mouse chromosome 7 in a region syntenic with the human chromosome 19q13.4 region that contains the *FcαR* gene (Alley et al. 1998; Kubagawa et al. 1997; Torkar et al. 1998; Yamashita et al. 1998a). This chromosome 19 region contains a monophyletic family of approximately 24 leukocyte receptor genes, many of which encode activating and inhibitory receptor pairs (Wilson et al. 2000). This region of the human genome is therefore referred to as the leukocyte receptor complex or cluster (LRC).

The closest human relatives of the *Pir* genes are the Ig-like transcripts/ leukocyte Ig-like receptors/monocyte Ig-like receptors (ILT/LIR/MIR) (Cosman et al. 1997; Samaridis and Colonna 1997; Wagtmann et al. 1997). The mouse PIR and human ILT/LIR/MIR families both contain activating and inhibitory members with similar EC regions. However, whereas the mouse PIRs have six Ig-like domains, the ILT/LIR/MIRs have four Ig-like domains. Another distinction is that mice possess a single inhibitory *Pirb* gene and multiple activating *Pira* genes, while humans have multiple inhibitory gene members and relatively few activating Ig-like receptor genes. The hallmark paired expression of PIR-A and PIR-B in mouse myeloid, B, and dendritic cell lineages is not a general rule for their human counterparts, and the cellular distribution of ILT/LIR/MIR expression extends to include subpopulations of the NK cells and T cells.

The human LRC includes other Ig-like receptor genes that are more distantly related to mouse PIRs. The natural killer cell Ig-like receptor (KIR) gene family is one of the best characterized LRC encoded gene families. With two or three Ig-like domains in their EC regions, the KIRs are restricted in their expression to NK and NK-T cells. Although distinct from the rodent PIRs and human ILT/LIR/MIRs, the KIRs are also encoded by a multigene family of activating and inhibitory isoforms. Because of their NK cell restricted expression and their MHC class I specificity, investigators have attempted to identify KIR homologs in mice, but without success. The recent characterization of the LRC containing genomic interval indicates a ratio of young Alu repeats (AluJ) to old Alu repeats (AluS) in the *KIR* gene region that suggests this subfamily appeared within the primate genome approximately 20 million years ago (Wilson et al. 2000). If the *Kir* genes were not present in the last common ancestor of rodents and primates, this explains why they have not been identified in mice.

Another PIR relative located within the LRC, a 46 kD receptor named NKp46, is specifically expressed in NK cells (see chapter 25). NKp46 possesses two Ig-like domains in its EC region and a transmembrane region endowed with a positively charged arginine residue. NKp46 associates with the ITAM containing adapter protein, CD3ζ, to function as an activating receptor. Upon interaction with its ligand, NKp46 triggers NK cells to kill target cells. Like the FcαR, NKp46 is encoded by a single gene and has limited, but discernable, similarity to other LRC encoded Ig-like receptors and the rodent PIRs. Interestingly, NKp46 is the only human LRC encoded gene with rodent orthologs, the mouse activating receptor-1 (MAR-1) and rat killer cell Ig-like receptor-1 (KILR-1) genes that are located in syntenic regions of their respective genomes.

Chickens PIR homologs

The chicken Ig-like receptors (CHIR)-A and CHIR-B were identified by searching the NCBI expressed sequence tag (EST) database for similarity with the PIR-B EC region sequence. Characterization of full-length sequences corresponding to ESTs identified in this search revealed a pair of protein products with the hallmark features of an activating and inhibitory receptor pair (Dennis et al. 2000). CHIR-A is an activating receptor candidate on the basis of a positively charged histidine residue in its transmembrane segment and a relatively short cytoplasmic tail. Conversely, CHIR-B can be identified as an inhibitory receptor by its nonpolar transmembrane segment and the two ITIMs on its cytoplasmic tail. A preliminary analysis of the CHIR repertoire indicates that CHIR-A and CHIR-B are members of a polymorphic multigene family that includes activating and inhibitory receptor isoforms. CHIR-A and CHIR-B transcripts are expressed in chicken B and T cell lines, suggesting that, like their mammalian counterparts, CHIR-A and CHIR-B may participate in the regulation of adaptive immunity.

PIR and Fc receptors are distant relatives

The Fc receptor (FcR) genes for IgG and IgE are encoded in syntenic regions of human and mouse chromosome 1 (Grundy et al. 1989; Huppi et al. 1988), whereas the FcαR is encoded on human chromosome 19. While IgG and IgE FcR homologs are broadly represented in the mammalian class, FcαR homologs have not been identified in non-primates. An earlier phylogenetic analysis of mammalian FcRs suggested that the closely related bovine Fcγ2R and human FcαR belong to a distinct class of mammalian FcRs that differs from the IgG and IgE FcR family (Zhang et al. 1995). The precise relationship between the FcR for IgG and IgE, the FcαR, and the rodent PIRs has been difficult to resolve due to their extensive sequence divergence.

The identification of PIR homologs in chickens extends the phylogenetic view of Ig-like receptor families that can regulate the activation status of hematopoietic cells (Dennis et al. 2000). Although the two EC domain structure of the CHIRs differs from that of the six EC domain containing PIRs, the CHIR Ig-like domains retain amino acid sequence motifs characteristic of the PIR family. Interestingly, use of the CHIR Ig-like domain amino acid sequences in iterative searches of Genebank sequences suggests that the Ig-like domains of CHIR are similar to those of both the PIR and FcR families (Fig. 1). The case for this relationship is strengthened by a structure-based comparison of the a PIR homolog, KIR, with an FcR family member, FcγRIIb. The KIR and FcγRIIb structures have similar arrangements of their EC domains, with the first domain being bent at an acute angle relative to the second domain. Notably, the respective bending angles for KIR and FcγRIIb differ by approximately $120°$ in that their first domains bend to opposite sides when the second domains are placed in the same orientation. KIR and FcγRIIb domains have 8-10 β strands that are similar in their arrangement to V-set and C-set domains. The V-type feature is in the first β strand, which possesses a cis-proline residue causing a split into A and A' strands. The A' strand pairs with the G strand forming a short segment of parallel β sheets in an otherwise antiparallel structure. The strand topology (C-C'-D/E) represents a C-type feature for the KIR and FcγRIIb domains. Amino acid sequence alignment of the KIR and FcγRIIb EC2 domains highlights their similar folding topologies, the major difference being a short D strand in KIR. When superimposed, KIR and FcγRIIb structures have a root mean square deviation of atomic radii of 1.3 Angstroms in a comparison including 69 α carbons per domain. Despite only 18% amino acid identity, these findings thus indicate that the Ig-like domains of KIR

12

and FcγRIIb are remarkable similar. Moreover, among Ig superfamily members, the KIR and FcγRIIb domains are more closely related to each other than to any other known structure. As structural representatives of the PIR and FcR gene families, the similarity between KIR and FcγRIIb provides strong support for the phylogenetic relationship between the PIR and FcR families.

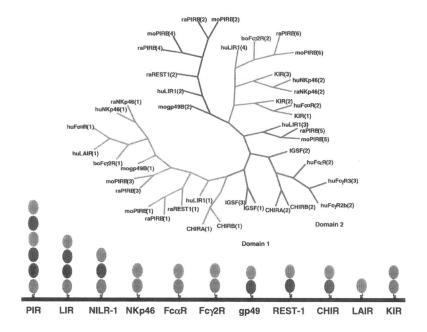

Fig. 1. Individual comparison of the Ig-like domains in currently identified PIR relatives. An all-against-all comparison was used to analyze potential relationships among the individual Ig-like domains of the PIR and FcR relatives. Individual domains are numbered from the amino terminus. Receptor names are indicated at the end of each branch and the domain being compared is indicated in parentheses. The tree depicts the relationship between PIR domains and the domains of PIR relatives, revealing similarity in domain organization among PIR-related gene families. The tree clusters the CHIR domain 1 (D1) with the PIR family (orange) and the CHIR D2 with the FcR family (red). The major branches of the tree are color coordinated with the Ig-like domains (solid ovals) of the illustrated receptors.

Concluding remarks

The proposed phylogenetic relationship between the PIR related gene families, including those encoded by the human LRC, and the FcR family predicts that an ancient gene family of activating and inhibitory Ig-like receptors has evolved into distinct PIR-like and FcR-like genetic lineages prior to the divergence of birds and mammals. This proposed genealogical relationship accords with the common functional strategies employed by the PIR-related gene families and the FcR gene families. The full spectrum of their ligands remains to be determined, but it is known that members of both the PIR and FcR families use their Ig-like domains for recognition of other Ig superfamily members and that receptor ligation initiated signaling cascades via either their cytoplasmic ITIMs or associated ITAM containing adapter proteins. The ligands so far identified for members of these related families are components of the immune system, namely MHC class I or class I-like antigens and immunoglobulin Fc regions. The nature of these ligands and what we know about the PIR and FcR families suggests that these receptor families arrived at their current form through coevolution with components of the adaptive immune system since its invention more than 450 million years ago. The conserved Ig domain is also seen in cell surface receptors of invertebrates, including the marine sponge *Geodia cydomium*, a representative of the oldest living metazoans (Schacke et al. 1994). Two classes of *Geodia* receptors containing Ig-like domains, receptor tyrosine kinases (RTK) and sponge cell adhesion molecules (SAM), have been identified (Blumbach et al. 1999). Each Ig-like *Geodia* receptor is a member of distinctive RTK or SAM multigene families. The polymorphic EC regions of both types of receptors include two Ig-like domains. The Ig-like domains in these receptors may represent the basic building blocks not only for the vertebrate immune system, but for all Ig-like receptor gene families. The increase in RTK and SAM expression observed during sponge allograft rejection suggests their involvement in self/non-self recognition. Metazoan species thus may possess the basic structural elements of the innate and adaptive immune systems. The formidable task remains of piecing together an evolutionary history from the phylogenetic footprints in extant versions of the Ig superfamily.

References

Alley TL, Cooper MD, Chen M, Kubagawa H (1998) Genomic structure of PIR-B, the inhibitory member of the paired immunoglobulin-like receptor genes in mice. Tissue Antigens 51;224-231

Bléry M, Kubagawa H, Chen CC, Vély F, Cooper MD, Vivier E (1998) The paired Ig-like receptor PIR-B is an inhibitory receptor that recruits the protein-tyrosine phosphatase SHP-1. Proc Natl Acad Sci USA 95:2446-2451

Blumbach B, Diehl-Seifert B, Seack J, Steffen R, Muller IM, Muller WE (1999) Cloning and expression of new receptors belonging to the immunoglobulin superfamily from the marine sponge Geodia cydonium. Immunogenetics 49:751-763

Chen CC, Hurez V, Brockenbrough JS, Kubagawa H, Cooper MD (1999) Paternal monoallelic expression of the paired immunoglobulin-like receptors PIR-A and PIR-B. Proc Natl Acad Sci USA 96:6868-6872

Cosman D, Fanger N, Borges L, Kubin M, Chin W, Peterson L, Hsu ML (1997) A novel immunoglobulin superfamily receptor for cellular and viral MHC class I molecules. Immunity 7:273-282

Dennis GJr, Kubagawa H, Cooper MD (2000) Paired Ig-like receptor homologs in birds and mammals share a common ancestor with mammalian Fc receptors [In Process Citation]. Proc Natl Acad Sci USA 97:13245-13250

Dennis G, Jr., Stephan RP, Kubagawa H, Cooper MD (1999) Characterization of paired Ig-like receptors in rats. J Immunol 163:6371-6377

Grundy HO, Peltz G, Moore KW, Golbus MS, Jackson LG, Lebo RV (1989) The polymorphic Fc gamma receptor II gene maps to human chromosome 1q. Immunogenetics 29:331-339

Hayami K, Fukuta D, Nishikawa Y, Yamashita Y, Inui M, Ohyama Y, Hikida M, Ohmori H, Takai T (1997) Molecular cloning of a novel murine cell-surface glycoprotein homologous to killer cell inhibitory receptors. J Biol Chem 272:7320-7327

Ho LH, Uehara T, Chen CC, Kubagawa H, Cooper MD (1999) Constitutive tyrosine phosphorylation of the inhibitory paired Ig-like receptor PIR-B. Proc Natl Acad Sci USA 96:15086-15090

Huppi K, Mock BA, Hilgers J, Kochan J, Kinet JP (1988) Receptors for Fc epsilon and Fc gamma are linked on mouse chromosome 1. J Immunol 141:2807-2810

Kubagawa H, Burrows PD, Cooper MD (1997) A novel pair of immunoglobulin-like receptors expressed by B cells and myeloid cells [see comments]. Proc Natl Acad Sci USA 94:5261-5266

Kubagawa H, Chen CC, Ho LH, Shimada TS, Gartland L, Mashburn C, Uehara T, Ravetch JV, Cooper MD (1999) Biochemical nature and cellular distribution of the paired immunoglobulin-like receptors, PIR-A and PIR-B. J Exp Med 189:309-318

Maeda A, Kurosaki M, Kurosaki T (1998a) Paired immunoglobulin-like receptor (PIR)-A is involved in activating mast cells through its association with Fc receptor gamma chain. J Exp Med 188:991-995

Maeda A, Kurosaki M, Ono M, Takai T, Kurosaki T (1998b) Requirement of SH2-containing protein tyrosine phosphatases SHP-1 and SHP-2 for paired immunoglobulin-like receptor B (PIR-B)-mediated inhibitory signal. J Exp Med 187:1355-1360

Maeda A, Scharenberg AM, Tsukada S, Bolen JB, Kinet JP, Kurosaki T (1999) Paired immunoglobulin-like receptor B (PIR-B) inhibits BCR-induced activation of Syk and Btk by SHP-1. Oncogene 18:2291-2297

Maliszewski CR, March CJ, Schoenborn MA, Gimpel S, Shen L (1990) Expression cloning of a human Fc receptor for IgA. J Exp Med 172:1665-1672

Morton HC, van Egmond M, van de Winkel JGJ (1996) Structure and function of human IgA Fc receptors (FcαR). Crit Rev Immunol 16:423-440

Ono M, Yuasa T, Ra C, Takai T (1999) Stimulatory function of paired immunoglobulin-like receptor-A in mast cell line by associating with subunits common to Fc receptors. J Biol Chem 274:30288-30296

Samaridis J, Colonna M (1997) Cloning of novel immunoglobulin superfamily receptors expressed on human myeloid and lymphoid cells: structural evidence for new stimulatory and inhibitory pathways. Eur J Immunol 27:660-665

Schacke H, Rinkevich B, Gamulin V, Muller IM, Muller WE (1994) Immunoglobulin-like domain is present in the extracellular part of the receptor tyrosine kinase from the marine sponge Geodia cydonium. J Mol Recognit 7:273-276

Torkar M, Norgate Z, Colonna M, Trowsdale J, Wilson MJ (1998) Isotypic variation of novel immunoglobulin-like transcript/killer cell inhibitory receptor loci in the leukocyte receptor complex. Eur J Immunol 28:3959-3967

Wagtmann N, Rojo S, Eichler E, Mohrenweiser H, Long EO (1997) A new human gene complex encoding the killer cell inhibitory receptors and related monocyte/macrophage receptors. Curr Biol 7:615-618

Wilson MJ, Torkar M, Haude A, Milne S, Jones T, Sheer D, Beck S, Trowsdale J (2000) Plasticity in the organization and sequences of human KIR/ILT gene families. Proc Natl Acad Sci USA 97:4778-4783.

Yamashita Y, Fukuta D, Tsuji A, Nagabukuro A, Matsuda Y, Nishikawa Y, Ohyama Y, Ohmori H, Ono M, Takai T (1998a) Genomic structures and chromosomal location of p91, a novel murine regulatory receptor family. J Biochem (Tokyo) 123:358-368

Yamashita Y, Ono M, Takai T (1998b) Inhibitory and stimulatory functions of paired Ig-like receptor (PIR) family in RBL-2H3 cells. J Immunol 161:4042-4047

Zhang G, Young JR, Tregaskes CA, Sopp P, Howard CJ (1995) Identification of a novel class of mammalian Fc gamma receptor. J Immunol 155:1534-1541

3
Genomic organization of the ILT11 gene, a novel member of the Leukocyte Receptor Cluster (LRC)

Hagen Wende, Andreas Ziegler and Armin Volz

Institut für Immungenetik, Universitätsklinikum Charité, Humboldt-Universität zu Berlin, Spandauer Damm 130, D-14050 Berlin, Germany

Summary. The chromosomal region 19q13.4 harbors the human leukocyte receptor cluster (LRC) which has been demonstrated to contain over 25 genes encoding leukocyte-expressed receptors of the immunoglobulin superfamily. Among these are the killer cell inhibitory receptors (KIR), immunoglobulin like transcripts (ILT), leukocyte associated inhibitory receptors (LAIR), the Fc receptor for IgA (FCAR) as well as the gene for NCR1. ILTs are a family of homologous receptors with two or four immunoglobulin domains. We have determined the genomic organization of ILT11, a novel member of this family that comprises only two immunoglobulin domains. These are most homologous to the first and second immunoglobulin domain of four domain ILTs. ILT11 has a short cytoplasmic tail and a charged amino acid in its transmembrane region, which is characteristic of an activatory receptor. Additionally, the translation starts at a different codon, if compared to other family members.

Key words. Immunoglobulin like transcript, ILT, Leukocyte receptor cluster, LRC, Chromosome 19, NK cell receptor

Introduction

Target cell recognition by natural killer (NK) cells is mediated by inhibitory and stimulatory receptors. These receptors enable NK cells to monitor HLA class I expression on target cells (Yokoyama 1998). The HLA class I-specific inhibitory receptors found to be responsible for this effect belong to two structurally distinct families: (1) the heterodimeric CD94/NKG2 receptors are members of the C-type-like lectin family and recognize HLA-E assembled with the leader sequence of certain HLA class I molecules (Borrego et al. 1998); (2) the killer cell inhibitory receptors (KIR) (Colonna and Samaridis 1995) are highly polymorphic glycoproteins belonging to the immunoglobulin superfamily (Ig-SF). They recognize polymorphic regions on HLA-A, -B and -C molecules (reviewed in Lanier 1998).

Recently, a new family of KIR-related receptors has been discovered and termed immunoglobulin-like transcripts (ILT) (Samaridis and Colonna 1997), leukocyte immunoglobulin-like receptors (LIR) (Cosman et al. 1997) or monocyte/macrophage inhibitory receptors (MIR) (Wagtmann et al. 1997). These receptors are encoded together with KIR, LAIR, FCAR and NCR1 in the leukocyte receptor cluster (LRC) on chromosome 19q13.4 (Wagtmann et al. 1997; Wilson et al. 2000; Wende et al. 2000).

ILTs are preferentially expressed on monocytes, macrophages, dendritic cells and granulocytes. They are characterized by two or four Ig-SF domains. ILTs have either a long cytoplasmic tail containing immunoreceptor tyrosine based inhibitory motifs (ITIM) or a short cytoplasmic tail with a charged arginine residue in the transmembrane region. They mediate inhibitory or activatory signals, respectively. ILT2 and ILT4 have also been shown to interact with HLA class I molecules, while the ligands of the other ILTs remain elusive (reviewed in Colonna et al. 1999).

We here describe a full size cDNA and the genomic organization of a novel member of this family, called ILT11 (Wende et al. 2000). This new receptor has two Ig-SF domains and a short cytoplasmic tail. Its 7 exons span approximately 6 kb genomic DNA. The start codon used by all other family members is missing in the ILT11 gene, leading to a translation start that differs from all other ILTs. Furthermore, ILT11 is characterized by a charged amino acid in its transmembrane region, making it likely that the ILT11 product is an activatory receptor.

Materials and methods

PCR and Cloning of PCR Fragments

Fragments were PCR-amplified in a total volume of 30 μl using 10 pmol of each primer and 0.6 U Taq (Perkin Elmer) per reaction. The PCR protocol comprised: 94°C 30 sec, annealing temp 20 sec, 72°C 1 min for 30 cycles, final extension 7 min at 72°C.

Amplifications of large PCR fragments were performed using 0.5 U TaKaRa LA Taq according to manufacturer's instructions and as described above. PCR conditions were as follows: 94°C 30 sec, annealing Temp 20 sec, 68°C 10 min for 25 cycles, final extension 15 min at 68°C.

If necessary, PCR products were cloned into pCRII using the TOPO TA cloning kit (Invitrogen) according to the manufacturer's instructions. Recombinant clones were identified by blue/white screening and used for sequencing or insert isolation.

DNA Sequencing

Plasmid DNA was isolated by a standard alkaline lysis procedure and then directly used for sequencing. Cycle sequencing was performed using the Thermo Sequenase Kit (Amersham Pharmacia Biotech) with labeled primers. Reaction products were analyzed on a model 4000 automated sequencer (Licor).

Table 1. Primer sequences used for PCR

Primer name	Sequence	Direction
GSP1-ILT11	GGGCAGGTATGCAGCTCAGCCTGGGCT	3'
GSP2-ILT11	TCCTCCATGGATCACCCC	5'
ILT11-IgD1→5'	ACCGGATGGTCACAGAGT	5'
ILT11-IgD2-Rv	GGTTCTGACCATACCTGCA	5'
Primers used in the vectorette approach		
ILT11-IgD2→3'	TATGGCTCTCGCAGGCAT	3'
ILT11-IgD1→5'	ACCGGATGGTCACAGAGT	5'

Sequence analysis

Sequences were analyzed using the Blast algorithm (Altschul et al. 1997) to search for homologous sequences. Multiple sequence alignments were performed using Multalin (Corpet 1988). Transmembrane region prediction was done using TMAP (Persson and Argos 1996; Persson and Argos 1994).

Results

Cloning of ILT11 full-size cDNA

To obtain a full size cDNA of ILT11, we performed 5' rapid amplification of cDNA ends (RACE) on cDNA prepared from peripheral blood leukocytes using the Marathon™ cDNA Amplification KIT. The Primer ILT11-IgD1→5' was used to amplify the 5' end of the ILT11 cDNA. A ~350bp PCR fragment was obtained and sequenced following cloning into pCRII. The obtained sequence was then used to design an ILT11-specific primer fitting the most 5' part of ILT11. To amplify the full size cDNA, this primer (GSP1-ILT11) was then used together with GSP2-ILT11, which is based on ILT11 sequence published previously (Wende et al. 2000). The PCR resulted in only one strong band of ~1.4 kb. This PCR product was cloned and sequenced (Acc. No.: AF324830). The ILT11 cDNA contains a potential open reading frame (ORF) encoding 299 amino acids (AA). The resulting protein is expected to have a mass of 33 kDa. It contains two Ig-like domains in its extracellular portion, a charged amino acid (Arg) in the trans-membrane region and a short cytoplasmic tail. Interestingly, the start codon found in all other ILTs was not present in ILT11. However, two different potential translation start sites are present in ILT11, the first 18 AA upstream and the second 5 AA downstream of the expected methionine codon found in all other ILTs (Fig. 1).

Genomic organization of ILT11

The genomic organization of ILT11 was analyzed by performing PCR on PAC LLNLP704J23750, which has been shown to contain ILT11 (Wende et al. 2000). Using the primer GSP1-ILT11 at the 5' end of the gene and ILT11-IgD2-Rv complementary to the 3' end of the potential second Ig domain, we found a 1.7 kb PCR product which has been cloned and sequenced.

```
                 1          10         20         30         40
ILT11    MAPWSHPSAQLQPVGGDAVSPALMVLLCLGLSLGPRTHVQAGNLSKATLW
   ILT1                     MT.I.T..I...............H.P.P...
   ILT6                     MT.I.T..I......D........P.P.P...
   LIR6                     MT.IVT..I..R............T.P.P...
   ILT2                     MT.I.T..I...............H.P.P...
   ILT4                     MT.IVT..I.............T.TIP.P...
   ILT5                     MT...TA............RM...PFP.P...
   LIR8                     MTLT.S..I.....V....C....T.P.P...
   ILT7                     MT.I.T..I..........R...E..P.PI..
   ILT8                     MT.I.T..I...............PFP.P...
  ILT10                     MVSI.T.....R....QKAQAL..T.P.PS..
   ILT3                     MI.TFTA.............M...P.P.P...
Consensus    ..................mtpiltv.i..g....prthvq...lp.pt..
```

Fig. 1. Alignment of the 5' region of various ILT proteins with ILT11 showing the different start codon of ILT11

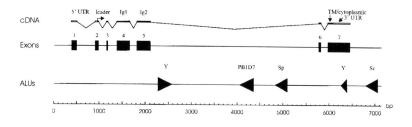

Fig. 2. Genomic organization of ILT11 with domain organization of the cDNA on top, exon organization below and occurrence of ALU elements at the bottom.

To determine the genomic sequences further 5' and 3', we used a modified protocol of the vectorette technique (Riley et al. 1990) on PAC LLNLP704J23750 with gene-specific primers. In total, 2371 bp were sequenced on PAC LLNLP704J23750 corresponding to basepair 1 to 2371 depicted in Fig. 2. This includes exon 1 to 5. The sequence further 3' was taken from PAC LLNLP704H20598, which is currently sequenced at the Sanger Center (ftp://ftp.sanger.ac.uk/pub/human/sequences/Chr_19/unfinished_sequence/).

Interestingly, PAC LLNLP704H20598 was previously found to be negative for ILT11. This may reflect the instability of genomic clones observed in this region, resulting in clones containing different amounts of material (Wende et al. 2000).

The comparison of the cDNA sequence with the genomic sequence of PAC LLNLP704J23750 and PAC LLNLP704H20598 led to the genomic organization of ILT11 shown in Figs. 2 and 3:

The ILT11 gene is composed of seven exons including a 5' untranslated exon, two exons encoding the potential leader, one exon for either Ig domain, a very small exon for the stalk region and a large exon containing the transmembrane and cytoplasmic domains. The intron between exon 5 and 6 does not contain any further pseudo-exons or sequences homologous to the 3rd and 4th Ig domains of other ILTs. This is a unique feature among the ILTs analyzed so far. Instead, this intron contains 3 repetitive elements.

	Exon 1 (123 bp) 5'UTR	TGACCATGgtaaggac...	Intron 1 395 bp	...
...cctgccagGCACCATG	Exon 2 (85 bp) Start/leader	CTGCCTCGgtgagatt...	Intron 2 166 bp	...
...tcttccagGGCTGAGT	Exon 3 (35 bp) leader	GCAGGCAGgtgagtct...	Intron 3 186 bp	...
...ccttccagGGAACCTC	Exon 4 (285 bp) Ig Domain 1	GGTGACAGgtgagagg...	Intron 4 147 bp	...
...tctcctagGATTCTAC	Exon 5 (303bp) Ig Domain 2	GGTCTCAGgtgaggaa...	Intron 5 3653 bp	...
...attttcagGAGCAGCT	Exon 6 (51 bp) stalk	TGGGACTGgtgagtga...	Intron 6 148 bp	...
...atccccagCCTCACAC	Exon 7 (480 bp) TM/Cytoplasmic domain/3'UTR	aaaaaaaaaaaaaa		

Fig. 3. Intron/exon boundaries of ILT11 with splice donor/acceptor motifs in bold, all following the AG/GT rule. Exon sequences are capitalized and sizes of exons are given in brackets.

Discussion

The leukocyte receptor cluster (LRC) on chromosome 19q13.4 has been shown to contain 11 ILT genes in two clusters. The gene content and evolutionary relationships between the ILT genes in these clusters have been studied in detail (Wagtmann et al. 1997; Wilson et al. 2000; Wende et al. 2000). Our studies resulted in the identification of ILT11, a novel member of the ILT family. A comparison of the cDNA sequence (Wende et al. 2000) and the corresponding genomic region of PACs LLNLP704J23750 and LLNLP704H20598 revealed that exons 2-4 and part of exon 5 of ILT11 were highly similar to the organization and sequence of ILT3 (Torkar et al. 1998), the only other ILT with two Ig domains. The potential leader is encoded by two exons and each Ig domain is encoded by a separate exon, which is typical for Ig-SF members. However, as shown for the LIR6 and ILT2 gene (Wilson et al. 2000), ILT11 has an additional exon 5' of the predicted translation start site, which was not found in the ILT3 gene. The relatively small exon 6 encoding the stalk is followed by exon 7 corresponding to the transmembrane region and the cytoplasmic domain. This organization together with the arginine in the transmembrane domain is characteristic of an activating receptor. The ILT11 gene does not contain pseudo-exons or fragments of exons that potentially code for the third and fourth Ig domain. Instead, the intron between exon 5 and 6 was found to be filled with three repeat elements, which is very unusual among ILT genes, as shown in the ILTtel cluster (Wilson et al. 2000). Therefore, we suppose that ILT11 arose from a 4 domain ILT by deletion or replacement of Ig domains 3 and 4.

This is in contrast to the situation found in ILT3, the only other ILT with two Ig domains. A high sequence similarity between ILT3 and ILT11 was found on the genomic level between exon 2 and the first 45 base pairs of exon 5 of ILT11, the latter corresponding to Ig domain 2. However, the second part of Ig domain 2 of ILT3 is highly homologous to the fourth Ig domain of other ILTs and probably resulted from a deletion of the intervening genomic segment, containing the second part of Ig domain two, all of Ig domain three and the first 45 base pairs of Ig domain four.

The start codon of ILT11 was not found at the expected position. Instead, two alternative start sites are present. The first was found upstream and would add 18 AA to the leader. On the hydrophobicity blot, at least the first 10 AA are not hydrophobic. The second start site was found downstream and reduces the leader by five AA. It is therefore unclear whether the translation product of ILT11 includes a functional leader sequence and results in a surface-expressed protein.

Altogether, the functional relevance, specificity and precise expression pattern of ILT11 remains to be determined.

Acknowledgements

We are very grateful for the expert technical assistance of Ms. Katja Laun and Waltraud Bangel. This work was supported by the European Union through grant BMH4-CT96-1105 (to A.Z.), the Sonnenfeld-Stiftung (Berlin) and the Monika Kutzner Stiftung (Berlin). H.W. receives a scholarship from the Sonnenfeld-Stiftung (Berlin).

References

Altschul SF, Madden TL, Schaffer AA, Zhang J, Zhang Z, Miller W, Lipman DJ (1997) Gapped BLAST and PSI-BLAST: a new generation of protein database search programs. Nucleic Acids Res 25:3389-3402

Borrego F, Ulbrecht M, Weiss EH, Coligan JE, Brooks AG (1998) Recognition of human histocompatibility leukocyte antigen (HLA)-E complexed with HLA class I signal sequence-derived peptides by CD94/NKG2 confers protection from natural killer cell-mediated lysis. J Exp Med 187:813-818

Colonna M, Nakajima H, Navarro F, Lopez-Botet M (1999) A novel family of Ig-like receptors for HLA class I molecules that modulate function of lymphoid and myeloid cells. J Leukoc Biol 66:375-381

Colonna M, Samaridis J (1995) Cloning of immunoglobulin-superfamily members associated with HLA-C and HLA-B recognition by human natural killer cells [see comments]. Science 268:405-408

Corpet F (1988) Multiple sequence alignment with hierarchical clustering. Nucleic Acids Res 16:10881-10890

Cosman D, Fanger N, Borges L, Kubin M, Chin W, Peterson L, Hsu ML (1997) A novel immunoglobulin superfamily receptor for cellular and viral MHC class I molecules. Immunity 7:273-282

Lanier LL (1998) Follow the leader: NK cell receptors for classical and nonclassical MHC class I. Cell 92:705-707

Persson B, Argos P (1994) Prediction of transmembrane segments in proteins utilising multiple sequence alignments. J Mol Biol 237:182-192

Persson B, Argos P (1996) Topology prediction of membrane proteins. Protein Sci 5:363-371

Riley J, Butler R, Ogilvie D, Finniear R, Jenner D, Powell S, Anand R, Smith JC, Markham AF (1990) A novel, rapid method for the isolation of terminal sequences from yeast artificial chromosome (YAC) clones. Nucleic Acids Res 18:2887-2890

Samaridis J, Colonna M (1997) Cloning of novel immunoglobulin superfamily receptors expressed on human myeloid and lymphoid cells: structural evidence for new stimulatory and inhibitory pathways. Eur J Immunol 27:660-665

Torkar M, Norgate Z, Colonna M, Trowsdale J, Wilson MJ (1998) Isotypic variation of novel immunoglobulin-like transcript/killer cell inhibitory receptor loci in the leukocyte receptor complex. Eur J Immunol 28:3959-3967

Wagtmann N, Rojo S, Eichler E, Mohrenweiser H, Long EO (1997) A new human gene complex encoding the killer cell inhibitory receptors and related monocyte/macrophage receptors. Curr Biol 7:615-618

Wagtmann N, Rojo S, Eichler E, Mohrenweiser H, Long EO (1997) A new human gene complex encoding the killer cell inhibitory receptors and related monocyte/macrophage receptors. Curr Biol 7:615-618

Wende H, Volz A, Ziegler A (2000) Extensive gene duplications and a large inversion characterize the human leukocyte receptor cluster. Immunogenetics 51:703-713

Wilson MJ, Torkar M, Haude A, Milne S, Jones T, Sheer D, Beck S, Trowsdale J (2000) Plasticity in the organization and sequences of human KIR/ILT gene families. Proc Natl Acad Sci USA 97:4778-4783

Yokoyama WM (1998) Natural killer cell receptors. Curr Opin Immunol 10:298-305

4
Regulated expression of non-polymorphic gp49 molecules on mouse natural killer cells

Lawrence L. Wang and Wayne M. Yokoyama

Howard Hughes Medical Institute, Division of Rheumatology, Departments of Medicine
and of Pathology and Immunology, Box 8045, Washington University School of Medicine,
St. Louis, MO 63110, USA

Summary. Murine natural killer (NK) cells express inhibitory receptors belonging
to the C-type lectin-like (Ly-49, CD94/NKG2) and Ig-superfamily related (gp49)
receptors. The murine gp49B receptor displays structural homology with human
killer inhibitory receptors (KIR) and other Ig-like receptors. By contrast to most of
these receptors, gp49 molecules are not expressed by resting NK cells but are
expressed when NK cells are activated in vitro. Importantly, gp49 molecules are
also upregulated in the context of murine cytomegalovirus infection. Furthermore,
the gp49 molecules do not appear to display allelic polymorphism, suggesting a
unique functional role for gp49 molecules in regulating NK cells long after initial
stimuli.

Key words. NK cells, gp49, Receptors, Expression

Abbreviations. NK, natural killer; KIR, killer inhibitory receptor; MCMV,
murine cytomegalovirus; ITIM, immunoreceptor tyrosine based inhibitory motif;
mAb, monoclonal antibody; LAK, IL-2-activated NK cell

Mouse and human natural killer (NK) cells are regulated by parallel receptor
systems. One of the first receptors to be characterized, Ly49A, was found to bind
MHC class I molecules and inhibit the capacity of NK cells to kill their targets
(Karlhofer et al. 1992). It is now clear that there is structural diversity in these
receptors (Yokoyama 1998). In mice, most of the defined NK cell receptors
belong to the C-type lectin superfamily. While it is still controversial whether
these receptors bind carbohydrates, they nevertheless bear a structural relationship
to molecules that bind carbohydrates in a Ca^{++}-dependent manner (Tormo et al.
1999), and they are expressed as disulfide-linked dimers with type II integral

membrane topology. Mouse NK cells recognize MHC class I with inhibitory Ly-49 and CD94/NKG2A receptors (Yokoyama 1998). The specificity of human NK cells is endowed not only by CD94/NKG2 receptors but also by killer inhibitory receptors (KIR), type I membrane proteins belonging to the Ig superfamily, which bind to specific HLA class I molecules (Long 1999). Despite their opposing membrane orientations, the inhibitory receptors mediate their effects through a conserved signaling mechanism: Recruitment of the SHP-1 tyrosine phosphatase to phosphorylated receptor immunoreceptor tyrosine-based inhibitory motifs (ITIMs). SHP-1 is thought to then dephosphorylate substrates that are involved in NK cell activation pathways. Multiple isoforms of either or both structural types of receptors may be expressed by a single NK cell. Importantly, studies to date indicate that the expression of such receptors is constitutive without significant alteration once the mature repertoire is established.

Whereas CD94/NKG2A orthologs are clearly present in both species (Ho et al. 1998; Lopez-Botet et al. 1997; Phillips et al. 1996; Vance et al. 1997), it is not yet clear if true orthologs of mouse Ly49 and human KIR exist in the reciprocal species. Despite their structural differences, mouse Ly49 and human KIR appear to subserve the same function, i.e., MHC class I-specific receptors that inhibit NK cell activation. A human Ly49 gene has been isolated but it appears to be a pseudogene (Westgaard et al. 1998) and functional human Ly49 molecules have yet to be described. Conversely, mouse NK cells express Ig-like receptors. The first recognized mouse NK cell receptor with structural homology to human KIR was the ITIM-containing gp49B receptor. This receptor was originally identified on mouse mast cells (Arm et al. 1991; Castells et al. 1994) and is also expressed on IL-2 activated NK (LAK) cells (Rojo et al. 1997; Wang et al. 1997). However, in contrast to C3H-derived mast cells, IL-2-activated NK cells from C57BL/6 mice express alternative gp49 transcripts (Fig. 1). cDNA cloning revealed 6 different clones, clone 10 and 11 were missing exons 3 and 4, giving rise to truncated gp49A; clone 6 represented gp49A missing half of exon 4; clone14 represented full length gp49B1. Two additional clones contained intronic sequences, probably representing splicing intermediates (not shown). It is unlikely that these results are due to strain differences since there is no evidence for allelic polymorphism of these genes (Wang et al. 1997; Wang and Yokoyama 1998). Like the NK cell inhibitory receptors belonging to either structural type, gp49B inhibits cellular activation by a mechanism involving the phosphorylation of its ITIM and recruitment and activation of SHP-1 (Lu-Kuo et al. 1999; Wang et al. 1999). The gp49A molecule has a much shorter cytoplasmic domain lacking ITIMs or any other tyrosine-based signaling motifs (Arm et al. 1991). The gp49A and B molecules have highly conserved extracellular domains with 89% amino acid identity. In this respect, gp49A and B resemble human KIRs which have ITIM-less non-inhibitory forms, termed killer activation receptors (KAR). Even though there are similarities between gp49 molecules human KIR/KAR, the expression of mouse gp49 is not restricted to NK cells. Moreover, even though gp49A appears to function as an activation receptor (Lee et al. 2000), it does not

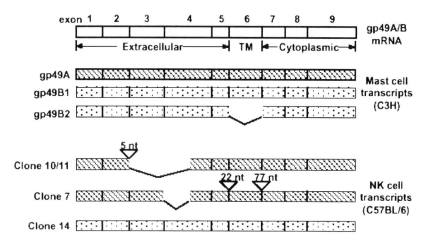

Fig. 1. Schematic representation of gp49 transcripts from C3H-derived mast cells (Arm et al. 1991; Castells et al. 1994) and IL-2-activated NK cells from C57BL/6 mice (Wang et al. 1997). Two additional gp49A transcripts were found in NK cells but these contained intronic sequences, probably representing splicing intermediates.

contain transmembrane charged residues that are utilized by ITIM-less KAR molecules to associate with signaling chains. These dissimilarities indicate that mouse gp49 molecules are not true orthologs of human KIR/KAR.

Other Ig-like receptors may be more closely related to the mouse gp49 receptors. The recently identified human ILT/LIR receptors (Borges et al. 1999; Colonna et al. 1999) may represent functional orthologs of mouse gp49. There are several known ILT/LIR family members that display heterogeneous tissue distribution. A few members such as ILT1 (LIR-7), and ILT2 (LIR-1) can be found on NK cells but also display tropism for cells of the myeloid lineage. ILT2 (LIR-1) is also found on B cells and T cell subsets. Like the gp49 family, there exist ITIM-less ILT/LIR molecules. Unlike gp49A, one such receptor, ILT1, through a charged residue in its transmembrane domain, can pair with FcεR and mediate activation of monocytes (Nakajima et al. 1999). There is also significant structural homology of mouse gp49 molecules with mouse PIR (Hayami et al. 1997; Kubagawa et al. 1997) and human ILT/LIR receptors. However, PIRs are generally not found on NK cells but instead are expressed on myeloid lineage and B cells (Kubagawa et al. 1999), unlike gp49. Thus, it is not yet clear how gp49 relates to these otherwise homologous receptors.

Nevertheless, gp49B can inhibit cellular activation pathways. When gp49B is co-crosslinked with FcεRI, it inhibits granule release from bone marrow-derived mast cells (Katz et al. 1996). The function of native gp49B on NK cells was addressed by similar studies. Cross-linking of the NK1.1 molecule with the PK136 mAb leads to cytokine production and target killing (Karlhofer and Yokoyama 1991; Kim and Yokoyama 1998). However, when IL-2-activated NK cells are co-cross-linked with an anti-gp49-specific mAb, cytokine secretion is inhibited (Wang et al. 2000). Furthermore, a chimeric receptor containing the gp49B cytoplasmic tail can impair NK cell cytolytic function (Rojo et al. 1997). These data suggest that gp49B plays an important regulatory function on NK cells but the cellular distribution of the gp49 receptors on primary hematopoietic cells was not previously known since most published studies had utilized mast cells and NK cells which have been cultured in vitro with cytokines.

One distinguishing feature of gp49 molecules turned out to be their regulated expression on NK cells. In recent studies, we described the surprising finding that gp49 molecules are not constitutively expressed on resting NK cells even though they are expressed by freshly isolated myeloid cells (Wang et al. 2000). When NK cells are stimulated in vitro by IL-2, they express gp49 molecules, including the inhibitory form, gp49B and cross-linking of gp49 molecules leads to inhibition of stimulation. Most importantly, gp49 expression is also induced on NK cells upon murine cytomegalovirus infection in vivo. Upregulated expression was found in MCMV-infected mice with targeted deficiency in IL-12β or IFNγ, suggesting that neither IL-12 or IFNγ is directly involved in stimulating expression of gp49 in MCMV infection. The kinetics of expression is similar to expression of another activation induced molecule, CD69, but in contrast, expression of gp49 does not occur on B or T cells. RT-PCR analysis revealed that both gp49A and gp49B transcripts were induced in activated NK cells. Thus, whereas the expression of other NK cell and Ig-like receptors is stable, the expression on NK cells of gp49 is regulated by activation.

The regulated expression of gp49 molecules suggests an important physiological role for gp49 receptors in regulating NK cell activity following viral infection. Both gp49A and gp49B transcripts are induced by NK cell activation but by cDNA cloning, we previously did not find full length transcripts for gp49A in IL-2-activated NK cells (Fig. 1). While the function of gp49A is just beginning to be elucidated as an activation receptor (Lee et al. 2000), perhaps the role of gp49 receptors in MCMV infection can be best understood in the context of the better characterized inhibitory receptors on NK cells. Whereas certain Ly-49 NK cell subsets in the spleen are altered during MCMV infection (Tay et al. 1999) (A. Dokun and W.M. Yokoyama, manuscript in preparation), perhaps by redistribution of these subsets or selective outgrowth, the inducible expression of gp49 appears to be a general response to NK cell activation. Early in the course of *Listeria monocytogenes* infection, another pathogen which involves an NK cell response (Unanue 1997), all NK cells also inducibly express gp49 (Wang and Yokoyama, unpublished observations). In mice treated with poly-I:C, an inducer of type I interferons, gp49 expression by NK cells is also induced (data not shown). Based on our in vitro observation that stimulation of mature NK cells induces gp49, the enhanced expression of gp49 following MCMV infection strongly suggest that gp49 expression is induced upon physiological stimuli and that the role of these receptors is to modulate NK cell activity some time after initial activation. Thus, it is not surprising that NK cell development is normal in mice with targeted deficiency in gp49B but these mice may manifest abnormalities in innate defense as suggested by our studies and those of the Long group (Rojo et al. 2000).

The apparent function of gp49 molecules should be distinguished from the role of Ly-49 and other MHC class I inhibitory receptors that are thought to act by inhibiting NK cell activation at the time of initial stimulation. These receptors are constitutively and stably expressed on individual NK cells once the mature repertoire is established (Dorfman and Raulet 1998; Valiante et al. 1997). Perhaps the constitutive and stable expression of such receptors is not surprising since these MHC class I-specific receptors must act immediately to control untoward NK cell activation. Although the ligands for the gp49 receptors are just beginning to be understood (Castells et al. 2000), ongoing studies do not clearly implicate MHC class I ligands (Wang and Yokoyama, unpublished data). Furthermore, the gp49 molecules do not appear to display allelic polymorphism. Whereas restriction fragment length polymorphic variants are easily detected between different inbred mouse strains with a Ly-49 cDNA (Yokoyama et al. 1990), the gp49 genes do not display apparent polymorphism (Wang et al. 1997; Wang and Yokoyama 1998). The allelic polymorphism of Ly-49 molecules may be important in maintenance of a repertoire of receptors capable of interacting with highly polymorphic MHC class I ligands. On the other hand, a limited diversity of gp49 molecules implies a more generic role and a limited universe of ligands. This is consistent with the recently described interaction of gp49B1 with a non-polymorphic ligand, $\alpha v \beta 3$ integrin (Castells et al. 2000). Hence, gp49 receptors and their ligands may play critical roles in returning the activated NK cell population back to resting state, thus playing a potentially important role in NK cell homeostasis.

Nevertheless, our data also imply that it is worth considering if any member of the growing family of Ig-like receptors may have important functions on not only resting cells but also on activated cells. A number of these receptors may be induced upon cellular activation on NK and other hematopoietic cells. Such findings would also strongly suggest that the physiological functions of these receptors may be important as regulators of cellular activation, long after the cells are first stimulated, rather than at the time of initial activation as currently envisioned.

Acknowledgments

We gratefully acknowledge research support from the NIH and Howard Hughes Medical Institute.

References

Arm JP, Gurish MF, Reynolds DS, Scott HC, Gartner CS, Austen KF, Katz HR (1991) Molecular cloning of gp49, a cell-surface antigen that is preferentially expressed by mouse mast cell progenitors and is a new member of the immunoglobulin superfamily. J Biol Chem 266:15966-15973

Borges L, Fanger N, Cosman D (1999) Interactions of LIRs, a family of immunoreceptors expressed in myeloid and lymphoid cells, with viral and cellular MHC class I antigens. Curr Top Microbiol Immunol 244:123-136

Castells M, Klickstein LB, Austen KF, Katz HR (2000) The immunoglobulin (Ig)-like inhibitory receptor gp49B1 binds integrin $\alpha v \beta 3$. FASEB J 14:A1128

Castells MC, Wu X, Arm JP, Austen KF, Katz HR (1994) Cloning of the gp49B gene of the immunoglobulin superfamily and demonstration that one of its two products is an early-expressed mast cell surface protein originally described as gp49. J Biol Chem 269:8393-8401

Colonna M, Nakajima H, Navarro F, Lopez-Botet M (1999) A novel family of Ig-like receptors for HLA class I molecules that modulate function of lymphoid and myeloid cells. J Leukoc Biol 66:375-381

Dorfman JR, Raulet DH (1998) Acquisition of Ly49 receptor expression by developing natural killer cells. J Exp Med 187:609-618

Hayami K, Fukuta D, Nishikawa Y, Yamashita Y, Inui M, Ohyama Y, Hikida M, Ohmori H, Takai T (1997) Molecular cloning of a novel murine cell-surface glycoprotein homologous to killer cell inhibitory receptors. J Biol Chem 272:7320-7327

Ho EL, Heusel JW, Brown MG, Matsumoto K, Scalzo AA, Yokoyama WM (1998) Murine nkg2d and Cd94 are clustered within the natural killer complex and are expressed independently in natural killer cells. Proc Natl Acad Sci USA 95:6320-6325

Karlhofer FM, Ribaudo RK, Yokoyama WM (1992) MHC class I alloantigen specificity of Ly-49+ IL-2-activated natural killer cells. Nature 358:66-70

Karlhofer FM, Yokoyama WM (1991) Stimulation of murine natural killer (NK) cells by a monoclonal antibody specific for the NK1.1 antigen. IL-2-activated NK cells possess additional specific stimulation pathways. J Immunol 146:3662-3673

Katz HR, Vivier E, Castells MC, McCormick MJ, Chambers JM, Austen KF (1996) Mouse mast cell gp49B1 contains two immunoreceptor tyrosine-based inhibition motifs and suppresses mast cell activation when coligated with the high-affinity Fc receptor for IgE. Proc Natl Acad Sci USA 93:10809-10814

Kim S, Yokoyama WM (1998) NK cell granule exocytosis and cytokine production inhibited by Ly-49A engagement. Cell Immunol 183:106-112

Kubagawa H, Burrows PD, Cooper MD (1997) A novel pair of immunoglobulin-like receptors expressed by B cells and myeloid cells. Proc Natl Acad Sci USA 94:5261-5266

Kubagawa H, Chen CC, Ho LH, Shimada TS, Gartland L, Mashburn C, Uehara T, Ravetch JV, Cooper MD (1999) Biochemical nature and cellular distribution of the paired immunoglobulin-like receptors, PIR-A and PIR-B. J Exp Med 189:309-318

Lee KH, Ono M, Inui M, Yuasa T, Takai T (2000) Stimulatory function of gp49A, a murine Ig-like receptor, in rat basophilic leukemia cells J Immunol 165:4970-4977

Long EO (1999) Regulation of immune responses through inhibitory receptors. Annu Rev Immunol 17:875-904

Lopez-Botet M, Perezvillar JJ, Carretero M, Rodriguez A, Melero I, Bellon T, Llano M, Navarro F (1997) Structure and function of the CD94 C-type lectin receptor complex involved in recognition of HLA class I molecules. Immunol Rev 155:165-174

Lu-Kuo JM, Joyal DM, Austen KF, Katz HR (1999) gp49B1 inhibits IgE-initiated mast cell activation through both immunoreceptor tyrosine-based inhibitory motifs, recruitment of src homology 2 domain-containing phosphatase-1, and suppression of early and late calcium mobilization. J Biol Chem 274:5791-5796

Nakajima H, Samaridis J, Angman L, Colonna M (1999) Human myeloid cells express an activating ILT receptor (ILT1) that associates with Fc receptor gamma-chain. J Immunol 162:5-8

Phillips JH, Chang CW, Mattson J, Gumperz JE, Parham P, Lanier LL (1996) CD94 and a novel associated protein (94ap) form a NK cell receptor involved in the recognition of HLA-A, HLA-B, and HLA-C allotypes. Immunity 5:163-172

Rojo S, Burshtyn DN, Long EO, Wagtmann N (1997) Type I transmembrane receptor with inhibitory function in mouse mast cells and NK cells. J Immunol 158:9-12

Rojo S, Stebbins CC, Peterson ME, Dombrowicz D, Wagtmann N, and Long EO (2000) Natural killer cells and mast cells from gp49B null mutant mice are functional. Mol Cell Biol 20:7178-7182

Tay CH, Yu LY, Kumar V, Mason L, Ortaldo JR, Welsh RM (1999) The role of LY49 NK cell subsets in the regulation of murine cytomegalovirus infections. J Immunol 162:718-726

Tormo J, Natarajan K, Margulies DH, Mariuzza RA (1999) Crystal structure of a lectin-like natural killer cell receptor bound to its MHC class I ligand. Nature 402:623-631

Unanue ER (1997) Inter-relationship among macrophages, natural killer cells and neutrophils in early stages of Listeria resistance. Curr Opin Immunol 9:35-43

Valiante NM, Uhrberg M, Shilling HG, Lienertweidenbach K, Arnett KL, Dandrea A, Phillips JH, Lanier LL, Parham P (1997) Functionally and structurally distinct NK cell receptor repertoires in the peripheral blood of two human donors. Immunity 7:739-751

Vance RE, Tanamachi DM, Hanke T, Raulet DH (1997) Cloning of a mouse homolog Of CD94 extends the family of C-type lectins on murine natural killer cells. Eur J Immunol 27:3236-3241

Wang LL, Blasioli J, Plas DR, Thomas ML, Yokoyama WM (1999) Specificity of the SH2 domains of SHP-1 in the interaction with the immunoreceptor tyrosine-based inhibitory motif-bearing receptor gp49B. J Immunol 162:1318-1323

Wang LL, Chu DT, Dokun AO, Yokoyama WM (2000) Inducible expression of the gp49B inhibitory receptor on NK cells. J Immunol 164:5215-5220

Wang LL, Mehta IK, Leblanc PA, Yokoyama WM (1997) Cutting Edge: Mouse natural killer cells express GP49b1, a structural homologue of human killer inhibitory receptors. J Immunol 158:13-17

Wang LL, Yokoyama WM (1998) Regulation of mouse NK cells by structurally divergent inhibitory receptors. Curr Top Microbiol Immunol 230:3-13

Westgaard IH, Berg SF, Orstavik S, Fossum S, Dissen E (1998) Identification of a human member of the Ly-49 multigene family. Eur J Immunol 28:1839-1846

Yokoyama WM (1998) Natural killer cell receptors. Curr Opin Immunol 10:298-305

Yokoyama WM, Kehn PJ, Cohen DI, Shevach EM (1990) Chromosomal location of the Ly-49 (A1, YE1/48) multigene family. Genetic association with the NK 1.1 antigen. J Immunol 145:2353-2358

5
Regulation of human *FcαR* gene expression

Toshibumi Shimokawa[1,2,4], Toshinao Tsuge[1,3], and Chisei Ra[1,2,4]

[1]Allergy Research Center, [2]Departmernt of Immunology, and [3]Department of Nephrology, Juntendo University School of Medicine, 2-1-1 Hongo, Bunkyo-ku, Tokyo 113-8421, Japan
[4]CREST, Japan Science and Technology Corporation (JST), Kawaguchi-city, Japan

Summary. The Fc receptor for IgA (FcαR, CD89) is expressed exclusively on human phagocytic cells including monocytes/macrophages, neutrophils, and eosinophils, and is capable of triggering various IgA-mediated effector functions. Altered FcαR expression has been reported in several diseases such as IgA nephropathy (IgAN) and allergic diseases, suggesting a role for FcαR in pathogenesis of diseases. The description of a promoter that directs tissue- or cell type-specific transcription could offer new insights into the mechanisms underlying aberrant gene expression associated with diseases. Here we show that at least two transcription factors, C/EBPα and an Ets-like factor, are functionally important for myeloid-specific expression of the *FcαR* gene. In addition, the *FcαR* promoter could be transactivated by PU.1, an Ets protein family member which has been shown to regulate the promoters of many myeloid-specific genes. Furthermore, in the *FcαR* promoter region there are two novel functional polymorphisms (T to C transition) that are associated with IgAN. The information obtained in this study will facilitate further analyses of activation stimuli and transcription factors involved in FcαR-mediated immune system in various pathological situations of diseases.

Key words. FcαR, CD89, Promoter, Polymorphism, IgA nephropathy

Introduction

In human, IgA is the prominent immunoglobulin (Ig) in secretions. One of the protection mechanisms of IgA is implicated in IgA-mediated neutralization and removal of environmental antigens from mucosal sites (Underdown and Schiff 1986). On the other hand, earlier reports describing the Fc receptor for IgA also suggested an active mechanism involving interaction of IgA-immune complexes (IgA-IC) with the specific Fc receptor expressed on effector cells (Fanger et al.

1980; Gauldie et al. 1983). It has become almost certain since a decade ago the gene encoding a human myeloid Fc receptor for IgA (*FcαR*, *CD89*) was indeed isolated (Maliszewski et al. 1990). Unfortunately, no homologue of the human *FcαR* gene has been isolated from any other species including mouse so far. Therefore, whereas in vivo roles of Fc receptors for IgE and IgG have been extensively investigated by gene-knockout experiments in mice, almost all of previous data regarding FcαR functions are based on in vitro studies using myeloid cells.

Human FcαR is expressed in myeloid cell lineage associated with mucosal surfaces, including monocytes/macrophages, neutrophils, and eosinophils (Maliszewski et al. 1990; Monteiro et al. 1992; Monteiro et al. 1993). Through the receptor, IgA-IC is capable of triggering effector functions of these cells such as phagocytosis (Yeaman and Kerr 1987), antibody-dependent cell-mediated cytotoxicity (ADCC) (Shen and Fanger 1981), superoxide production (Gorter et al. 1986; Shen et al. 1989), and the release of inflammatory mediators (Ferreri et al. 1986). In addition, endotoxin and several inflammatory cytokines such as tumor necrosis factor α (TNFα) have been shown to augment FcαR-mediated effector functions, which is accompanied by increased FcαR expression (Gessl et al. 1994; Hostoffer et al. 1994; Shen et al. 1994). Therefore, although the in vivo role of FcαR still remains to be clarified, expression of FcαR appears to be tightly regulated in a reflection of its potential functions.

Altered FcαR expression has been reported in several diseases. Decreased expression of this receptor has been shown in diseases associated with an increase in serum IgA levels, including alcoholic liver cirrhosis (Silvain et al. 1995), human immunodeficiency virus (HIV) infection (Grossetête et al. 1995), and IgA nephropathy (IgAN) (Grossetête et al. 1998), although increased FcαR expression in IgAN was also described in earlier reports (Kashem et al. 1994; Monteiro et al. 1995; Kashem et al. 1996; Kashem et al. 1997). Conversely, FcαR expression has been shown to be elevated on eosinophils of allergic individuals (Monteiro et al. 1993).

Considering the expression patterns, analysis of regulatory mechanisms controlling *FcαR* gene expression in normal and pathogenic situations could provide the foundation for understanding the in vivo roles of this receptor. The human *FcαR* gene was mapped to chromosome 19q13.4 (Kremer et al. 1992), and the nucleotide sequence upstream of the *FcαR*-coding region and its transcription start sites were determined (de Wit et al. 1995). We recently identified the functional promoter for the *FcαR* gene, and described its novel functional polymorphisms (Shimokawa et al. 2000). In this chapter, our analysis of the *FcαR* promoter has been extended, and we also discuss the physiological and pathological significance of the *FcαR* promoter polymorphisms.

Regulation of cell type-specific *FcαR* gene expression by C/EBPα and an Ets-like factor

The lineage-specific gene expression appears to be determined by the combined action of multiple nuclear factors, including transcription factors that mediate the differentiation signals (Shivdasani and Orkin 1996; Tenen et al. 1997). For example, monocyte-specific gene expression and monocyte development have been shown to involve cell-specific transcription factors, such as PU.1 and CCAAT/enhancer-binding proteins (C/EBPs), and ubiquitously expressed transcription factors, such as AML1 and Sp1 (Tenen et al. 1997; Valledor et al. 1998). Identification of transcription factors regulating cell type-specific expression of the *FcαR* gene provides the foundation for further analyses of more complicated mechanisms controlling *FcαR* expression in response to environmental factors and in pathogenic situations. As the first approach to determine the involved transcription factors, we cloned 929 bp of the *FcαR* 5'-flanking region from U937 genomic DNA by polymerase chain reaction (PCR), and fused it to the firefly luciferase (*Luc*) reporter gene. Transient transfection assay using this construct revealed that the 929-bp fragment directed a promoter activity in monocytic U937 cell line but not in nonhematopoietic Hela cell line (Shimokawa et al. 2000). In addition, this promoter region was also active in THP-1 monocytic cell line but not in Daudi B and Jurkat T cell lines. These results indicate that the *FcαR* 5'-flanking region contains a cell type-specific promoter.

Deletion analysis in U937 cells revealed that the promoter region between positions +134 and +95, whereby +1 denotes the major transcription start site (de Wit et al. 1995; Shimokawa et al. 2000), is essential for promoter activity. We identified in this region two potential binding motifs for transcription factors, C/EBP family (Akira et al. 1990) and Ets protein family members (Zhang et al. 1993). Indeed, point mutations in these sequences resulted in considerable reduction in *FcαR* promoter activity, indicating that both the Ets- and C/EBP-binding motifs are functionally important.

Electrophoretic mobility shift assay (EMSA) using U937 nuclear extracts demonstrated complexes specifically binding to the functional *FcαR* C/EBP motif, and supershift EMSA using antibodies against C/EBP family members revealed that they contained predominantly C/EBPα. Although C/EBP family members regulate the terminal differentiation of adipocytes and hepatocytes, expression of C/EBPα in hematopoietic system is restricted to myeloid cells (Scott et al. 1992).

On the other hand, the *FcαR* Ets motif specifically bound a nuclear factor, which was competed by oligonucleotides containing consensus-binding sequences for known Ets protein family members including *Drosophila* E74 (Urness and Thummel 1990). Although their competition efficiencies varied, these results strongly suggest an Ets family member, but which of the members binds to this site remains to be determined. We detected this Ets-like binding activity in a wide variety of cell lines, but predominantly in cells of hematopoietic origin with being relatively abundant in Jurkat T cell line.

Taken together, our functional analysis indicates that $Fc\alpha R$ gene expression is regulated by at least two transcription factors, C/EBPα whose expression in hematopoietic system is restricted to myeloid cells, and an Ets-like factor which is detected ubiquitously but predominantly in cell lines of hematopoietic origin.

Transactivation of the *FcαR* promoter by PU.1 and potential of kidney mesangial cells for FcαR expression

Our second approach involves cotransfection experiments using expression vectors of candidate transcription factors. Previous reports have shown that myeloid gene expression of high- and low-affinity Fc receptors for IgG (FcγRI and FcγRIII, respectively) are regulated by PU.1 (Feinman et al. 1994; Perez et al. 1994), an Ets family member which is expressed predominantly in monocytes/macrophages, neutrophils, and B cells (Klemsz et al. 1990; Chen et al. 1995) and implicated as a major regulator of myeloid differentiation (Scott et al. 1994; Olson et al. 1995; Henkel et al. 1996). Thus we next examined whether expression of the *FcαR* gene is regulated by this transcription factor. When the expression vector of PU.1 was cotransfected with the *FcαR* promoter-*Luc* construct into Hela cells, *FcαR* promoter activity was transactivated by several folds (Fig. 1). This suggests that the *FcαR* gene expression could be regulated by PU.1, similar to other myeloid Fc receptor promoters. Deletion analysis revealed

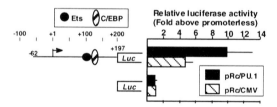

Fig. 1. Transactivation of the *FcαR* promoter by PU.1. Hela cells were transfected with the *Luc* reporter construct containing the *FcαR* promoter region between postions −62 and +197 or the promoterless vector in the presence of either expression vector of PU.1 (*pRc/PU.1*) or insertless vector (*pRc/CMV*). Luciferase activity was determined 24 h after transfection. Shown at the *left panel* is a schematic representation for the structure of the *FcαR* upstream-*Luc* and promoterless constructs. The major transcription start site is indicated as *bent arrow*. The *numbers* indicate nucleotide positions relative to the major transcription start site at +1. Values represent fold activity over the level of the promoterless construct along with pRc/CMV. *Thin bars* represent the standard error from five transfections.

that the transactivation by PU.1 required the *FcαR* promoter region between positions +3 and +58, but we found no canonical binding sequence for PU.1 within it.

We and others previously reported that FcαR is also expressed on human glomerular mesangial cells, suggesting a role for FcαR expressed on these cells in IgAN (Gómez-Guerrero et al. 1993; Bagheri et al. 1997; Suzuki et al. 1999). However, recent reports have described discrepant observations for mesangial expression of FcαR (Diven et al. 1998; Westerhuis et al. 1999; Leung et al. 2000), indicating that it still remains controversial. From the viewpoint of regulatory mechanisms underlying cell type-specific gene expression, future studies to determine whether mesangial cells express transcription factors required for *FcαR* gene expression may provide some explanation for such discrepancy. In this context, PU.1 has more recently been identified in rat mesangial cells (Harendza et al. 2000).

Identification and characterization of novel *FcαR* promoter polymorphisms and their association with IgAN

Our direct sequencing analyses of genomic DNA isolated from healthy Japanese individuals revealed that two nucleotides located at positions –114 and +56 are polymorphic (both are T to C transition) (Fig. 2). Because our third approach involves analyses for native mutations in promoter region, we fused the *FcαR* promoter region carrying each allele to the *Luc* reporter gene, and examined the effect of the mutations on *FcαR* promoter activity. Transfection assay of these

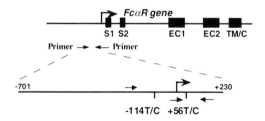

Fig. 2. Map of the oligonucleotide primers for PCR and DNA sequencing. The genomic organization of the human *FcαR* gene is shown at the *top panel*: Exons are presented by *filled boxes*; the signal peptide is encoded by exons *S1* and *S2*, the extracellular region is encoded by exons *EC1* and *EC2*, and exon *TM/C* encodes the transmembrane and cytoplasmic regions; the major transcription start site is indicated as *bent arrow* (de Wit et al. 1995). A DNA fragment spanning the *FcαR* upstream and S1-coding regions was amplified by PCR from genomic DNA, and subjected to direct DNA sequencing: The *numbers* indicate nucleotide positions relative to the major transcription start site at +1; polymorphisms examined in this work are indicated as *–114T/C* and *+56T/C*; Primers used for genomic PCR and direct sequencing are represented by *arrows* from their 5′ to 3′ ends.

38

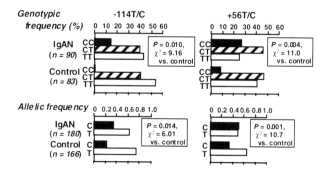

Fig. 3. Genotyping of promoter polymorphisms of the human $Fc\alpha R$ gene. A DNA fragment spanning the $Fc\alpha R$ upstream and S1-coding regions was amplified by PCR from genomic DNA of IgAN patients (*IgAN*) and healthy control (*Control*), and subjected to direct DNA sequencing. Genotypic and allelic frequencies were compared between groups using the chi-squared test. Differences were considered to significance at a value of $P < 0.05$. Each group was complied to the Hardy-Weinberg equilibrium.

constructs in monocytic U937 cell line revealed that nucleotide changes of T to C at both polymorphic sites significantly enhanced $Fc\alpha R$ promoter activity, strongly suggesting that the $Fc\alpha R$ promoter polymorphisms are physiologically functional (Shimokawa et al. 2000).

In EMSA using U937 nuclear extract, we detected a nuclear factor that binds to oligonucleotides carrying the –114C allele but not –114T allele-containing oligonucleotides. As far as we examined, the detection of this nuclear factor was restricted to cell lines of hematopoietic origin with being abundant in Daudi B cell line. On the other hand, detection of nuclear factor that binds to the +56 polymorphic site in an allele-specific manner has not been succeeded in our binding conditions.

Among diseases associated with altered $Fc\alpha R$ expression, IgAN has more increasing evidences for the involvement of $Fc\alpha R$ in its pathogenesis. This disease is a very common glomerulonephritis (GN), which is characterized by the depositions of IgA or IgA-IC in the glomerular mesangial areas, and in many cases the levels of serum IgA and IgA-IC are increased (Clarkson et al. 1977). Earlier reports described a correlation of IgAN or its severity with increased $Fc\alpha R$ expression and IgA-mediated superoxide generation (Kashem et al. 1994; Kashem et al. 1996; Monteiro et al. 1995; Kashem et al. 1997). However, Grossetête et al. (1998) have recently reported that $Fc\alpha R$ expression on monocytes and neutrophils of IgAN patients is decreased in association with elevated levels of serum IgA, and suggested that defective clearance of systemic IgA by $Fc\alpha R$ might contribute to mechanisms underlying augmented serum IgA levels in the patients. As a direct evidence for the importance of $Fc\alpha R$ in IgAN, Launay et al. (2000) have more recently reported that transgenic mice whose monocytes were enforced to express human FcαR came down with symptoms of IgAN.

Although the pathogenic mechanism of IgAN remains uncertain, it has been recognized for almost two decades that predisposing genetic factors determine susceptibility to IgAN (Schena et al. 1995). Because our in vitro data obtained by transfection assay suggest a functional role of the *FcαR* promoter polymorphisms in regulation of *FcαR* gene expression, we examined these polymorphisms in IgAN patients. Our direct sequencing analysis in Japanese population demonstrated that both C alleles at the –114 and +56 polymorphic sites significantly accumulate in IgAN patients compared with healthy controls (Fig. 3). While there are reports suggesting the importance of FcαR in IgAN, this is the first identification of structural change in the *FcαR* gene associated with IgAN.

IgAN is known to be a heterogeneous GN that shows various genotypes and phenotypes, and in fact many investigators have thus far described an association with various polymorphic genes including HLA antigen, complement, immunoglobulin, and T cell receptor genes (Schena et al. 1995). Moreover, it is thought that in combination with environmental risk factors different heritable factors simultaneously increase the development and pathogenesis of IgAN. Thus, how each genotype contributes to susceptibility to and/or the pathogenesis of IgAN allows us to determine how they interact, and will lead to determination of useful pre-clinical markers and to development of a practical application of molecular findings for a new intervention therapy. Regarding the *FcαR* promoter polymorphisms, we observed that *FcαR* promoter activity was enhanced by the C alleles at positions –114 and +56, which are associated with IgAN. Although this in vitro observation is consistent with earlier reports of increased FcαR expression (Kashem et al. 1994; Monteiro et al. 1995; Kashem et al. 1996; Kashem et al. 1997), decreased FcαR expression of circulating monocytes has recently been implicated in IgAN (Grossetête et al. 1998; Launay et al. 2000). In this context, a soluble form of FcαR has more recently been identified in IgAN patient serum as a putative pathogenic factor, and shown to be generated by proteolytic shedding in response to IgA (Launay et al. 2000). Therefore, decreased FcαR expression on monocytes could result from the release of soluble FcαR from the cell surface. The *FcαR* promoter polymorphisms may be involved in one of the mechanisms underlying susceptibility to and/or pathogenesis of IgAN through increasing *FcαR* gene expression, which might lead to increase soluble FcαR.

Acknowledgments

We thank Drs. Shoichi Suzuki and Michael J. Klemsz for providing expression vector of PU.1. We also thank Dr. Shinsaku Togo for experimental help. This work was supported in part by Grants in Aids from the Japanese Ministries of Education, Health and Welfare, and Environment Agency.

References

Akira S, Isshiki H, Sugita T, Tanabe O, Kinoshita S, Nishio Y, Nakajima T, Hirano T, Kishimoto T (1990) A nuclear factor for IL-6 expression (NF-IL6) is a member of a C/EBP family. EMBO J 9:1897-1906

Bagheri N, Chintalacharuvu SR, Emancipator SN (1997) Proinflammatory cytokines regulate FcαR expression by human mesangial cells in vitro. Clin Exp Immunol 107:404-409

Chen H-M, Zhang P, Voso MT, Hohaus S, Gonzalez DA, Glass CK, Zhang D-E, Tenen DG (1995) Neutrophils and monocytes express high levels of PU.1 (Spi-1) but not Spi-B. Blood 85:2918-2928

Clarkson AR, Seymour AE, Thompson AJ, Haynes WDG, Chan Y-L, Jackson B (1977) IgA nephropathy: A syndrome of uniform morphology, diverse clinical features and uncertain prognosis. Clin Nephrol 8:459-471

de Wit TPM, Morton HC, Capel PJA, van de Winkel JGJ (1995) Structure of the gene for the human myeloid IgA Fc receptor (CD89). J Immunol 155:1203-1209

Diven SC, Caflisch CR, Hammond DK, Weigel PH, Oka JA, Goldblum RM (1998) IgA induced activation of human mesangial cells: Independent of FcαR1 (CD 89). Kidney Int 54:837-847

Fanger MW, Shen L, Pugh J, Bernier GM (1980) Subpopulations of human peripheral granulocytes and monocytes express receptors for IgA. Proc Natl Acad Sci USA 77:3640-3644

Feinman R, Qiu WQ, Pearse RN, Nikolajczyk BS, Sen R, Sheffery M, Ravetch JV (1994) PU.1 and an HLH family member contribute to the myeloid-specific transcription of the FcγRIIIA promoter. EMBO J 13:3852-3860

Ferreri NR, Howland WC, Spiegelberg HL (1986) Release of leukotrienes C_4 and B_4 and prostaglandin E_2 from human monocytes stimulated with aggregated IgG, IgA, and IgE. J Immunol 136:4188-4193

Gauldie J, Richards C, Lamontagne L (1983) Fc receptors for IgA and other immunoglobulins on resident and activated alveolar macrophages. Mol Immunol 20:1029-1037

Gessl A, Willheim M, Spittler A, Agis H, Krugluger W, Boltz-Nitulescu G (1994) Influence of tumour necrosis factor-α on the expression of Fc IgG and IgA receptors, and other markers by cultured human blood monocytes and U937 cells. Scand J Immunol 39:151-156

Gómez-Guerrero C, González E, Egido J (1993) Evidence for a specific IgA receptor in rat and human mesangial cells. J Immunol 151:7172-7181

Gorter A, Hiemstra PS, Leijh PCJ, van der Sluys ME, van den Barselaar MT, van Es LA, Daha MR (1987) IgA- and secretory IgA-opsonized S. aureus induce a respiratory burst and phagocytosis by polymorphonuclear leucocytes. Immunology 61:303-309

Grossetête B, Viard J-P, Lehuen A, Bach J-F, Monteiro RC (1995) Impaired Fcα receptor expression is linked to increased immunoglobulin A levels and disease progression in HIV-1-infected patients. AIDS 9:229-234

Grossetête B, Launay P, Lehuen A, Jungers P, Bach J-F, Monteiro RC (1998) Down-regulation of Fcα receptors on blood cells of IgA nephropathy patients: Evidence for a negative regulatory role of serum IgA. Kidney Int 53:1321-1335

Harendza S, Lovett DH, Stahl RAK (2000) The hematopoietic transcription factor PU.1 represses gelatinase A transcription in glomerular mesangial cells. J Biol Chem 275:19552-19559

Henkel GW, McKercher SR, Yamamoto H, Anderson KL, Oshima RG, Maki RA (1996) PU.1 but not Ets-2 is essential for macrophage development from embryonic stem cells. Blood 88:2917-2926

Hostoffer RW, Krukovets I, Berger M (1994) Enhancement by tumor necrosis factor-α of Fcα receptor expression and IgA-mediated superoxide generation and killing of *Pseudomonas aeruginosa* by polymorphonuclear leukocytes. J Infect Dis 170:82-87

Kashem A, Endoh M, Nomoto Y, Sakai H, Nakazawa H (1994) FcαR expression on polymorphonuclear leukocyte and superoxide generation in IgA nephropathy. Kidney Int 45:868-875

Kashem A, Endoh M, Nomoto Y, Sakai H, Nakazawa H (1996) Monocyte superoxide generation and its IgA-receptor in IgA nephropathy. Clin Nephrol 45:1-9

Kashem A, Endoh M, Yano N, Yamauchi F, Nomoto Y, Sakai H, Kurokawa K (1997) Glomerular FcαR expression and disease activity in IgA nephropathy. Am J Kidney Dis 30:389-396

Klemsz MJ, McKercher SR, Celada A, van Beveren C, Maki RA (1990) The macrophage and B cell-specific transcription factor PU.1 is related to the *ets* oncogene. Cell 61:113-124

Kremer EJ, Kalatzis V, Baker E, Callen DF, Sutherland GR, Maliszewski CR (1992) The gene for the human IgA Fc receptor maps to 19q13.4. Hum Genet 89:107-108

Launay P, Grossetête B, Arcos-Fajardo M, Gaudin E, Torres SP, Beaudoin L, de Serre NP-M, Lehuen A, Monteiro RC (2000) Fcα receptor (CD89) mediates the development of immunoglobulin A (IgA) nephropathy (Berger's disease): Evidence for pathogenic soluble receptor-IgA complexes in patients and CD89 transgenic mice. J Exp Med 191:1999-2009

Leung JCK, Tsang AWL, Chan DTM, Lai KN (2000) Absence of CD89, polymeric immunoglobulin receptor, and asialoglycoprotein receptor on human mesangial cells. J Am Soc Nephrol 11:241-249

Maliszewski CR, March CJ, Schoenborn MA, Gimpel S, Shen L (1990) Expression cloning of a human Fc receptor for IgA. J Exp Med 172:1665-1672

Monteiro RC, Cooper MD, Kubagawa H (1992) Molecular heterogeneity of Fcα receptors detected by receptor-specific monoclonal antibodies. J Immunol 148:1764-1770

Monteiro RC, Hostoffer RW, Cooper MD, Bonner JR, Gartland GL, Kubagawa H (1993) Definition of immunoglobulin A receptors on eosinophils and their enhanced expression in allergic individuals. J Clin Invest 92:1681-1685

Monteiro RC, Grossetête B, Nguyen AT, Jungers P, Lehuen A (1995) Dysfunctions of Fcα receptors by blood phagocytic cells in IgA nephropathy. Contrib Nephrol 111:116-122

Olson MC, Scott EW, Hack AA, Su GH, Tenen DG, Singh H, Simon MC (1995) PU.1 is not essential for early myeloid gene expression but is required for terminal myeloid differentiation. Immunity 3:703-714

Perez C, Coeffier E, Moreau-Gachelin F, Wietzerbin J, Benech PD (1994) Involvement of the transcription factor PU.1/Spi-1 in myeloid cell-restricted expression of an interferon-inducible gene encoding the human high-affinity Fcγ receptor. Mol Cell Biol 14:5023-5031

Schena FP (1995) Immunogenetic aspects of primary IgA nephropathy. Kidney Int 48:1998-2013

Scott LM, Civin CI, Rorth P, Friedman AD (1992) A novel temporal expression pattern of three C/EBP family members in differentiating myelomonocytic cells. Blood 80:1725-1735

Scott EW, Simon MC, Anastasi J, Singh H (1994) Requirement of transcription factor PU.1 in the development of multiple hematopoietic lineages. Science 265:1573-1577

Shen L, Fanger MW (1981) Secretory IgA antibodies synergize with IgG in promoting ADCC by human polymorphonuclear cells, monocytes, and lymphocytes. Cell Immunol 59:75-81

Shen L, Lasser R, Fanger MW (1989) My43, a monoclonal antibody that reacts with human myeloid cells inhibits monocyte IgA binding and triggers function. J Immunol 143:4117-4122

Shen L, Collins JE, Schoenborn MA, Maliszewski CR (1994) Lipopolysaccharide and cytokine augmentation of human monocyte IgA receptor expression and function. J Immunol 152:4080-4086

Shimokawa T, Tsuge T, Okumura K, Ra C (2000) Identification and characterization of the promoter for the gene encoding the human myeloid IgA Fc receptor (FcαR, CD89). Immunogenetics 51:945-954

Shivdasani RA, Orkin SH (1996) The transcriptional control of hematopoiesis. Blood 87:4025-4039

Silvain C, Patry C, Launay P, Lehuen A, Monteiro RC (1995) Altered expression of monocyte IgA Fc receptors is associated with defective endocytosis in patients with alcoholic cirrhosis: Potential role for IFN-γ. J Immunol 155:1606-1618

Suzuki Y, Ra C, Saito K, Horikoshi S, Hasegawa S, Tsuge T, Okumura K, Tomino Y (1999) Expression and physical association of Fcα receptor and Fc receptor γ chain in human mesangial cells. Nephrol Dial Transplant 14:1117-1123

Tenen DG, Hromas R, Licht JD, Zhang D-E (1997) Transcription factors, normal myeloid development, and leukemia. Blood 90:489-519

Underdown BJ, Schiff JM (1986) Immunoglobulin A: Strategic defense initiative at the mucosal surface. Annu Rev Immunol 4:389-417

Urness LD, Thummel CS (1990) Molecular interactions within the ecdysone regulatory hierarchy: DNA binding properties of the *Droshophila* ecdysone-inducible E74A protein. Cell 63:47-61

Valledor AF, Borràs FE, Cullell-Young M, Celada A (1998) Transcription factors that regulate monocyte/macrophage differentiation. J Leukoc Biol 63:405-417

Westerhuis R, van Zandbergen G, Verhagen NA, Klar-Mohamad N, Daha MR, van Kooten C (1999) Human mesangial cells in culture and in kidney sections fail to express Fc alpha receptor (CD89). J Am Soc Nephrol 10:770-778

Yeaman GR, Kerr MA (1987) Opsonization of yeast by human serum IgA anti-mannan antibodies and phagocytosis by human polymorphonuclear leukocytes. Clin Exp Immunol 68:200-208

Zhang L, Lemarchandel V, Romeo P-H, Ben-David Y, Greer P, Bernstein A (1993) The *Fli-1* proto-oncogene, involved in erythroleukemia and Ewing's sarcoma, encodes a transcriptional activator with DNA-binding specificities distinct from other Ets family members. Oncogene 8:1621-1630

Part B

Structural Features of Receptor Protein

6
Molecular recognition by Ig-like receptors, KIRs and FcγRs

Katsumi Maenaka[1], P. Anton van der Merwe[3], David I. Stuart[2],
Peter Sondermann[4], and E. Yvonne Jones[2]

[1]Structural Biology Center, National Institute of Genetics, Mishima, Shizuoka 411-8540, Japan
[2]Division of Structural Biology, Wellcome Trust Centre for Human Genetics, University of Oxford, Roosevelt Drive, Headington, Oxford, OX3 7BN and [3]Sir William Dunn School of Pathology, University of Oxford, Oxford, OX1 3RE U.K.
[4]Max-Planck-Institut für Biochemie, Abteilung Strukturforschung, Am Klopferspitz 18a, D-82152 Martinsried, Germany

Summary. Structural studies by us and other groups have shown that Killer cell Ig-like receptors (KIRs) and Fcγ receptors (FcγR) have a similar, unique topology (intermediate between I set and C2 set). In order to gain further insight into molecular recognition by these receptors, we have used surface plasmon resonance (SPR) to analyze the kinetic and thermodynamic properties of their interactions with their natural ligands.

A repertoire of KIRs with two or three tandem Ig domains in their extracellular regions is expressed on human natural killer (NK) cells. These KIRs activate or inhibit NK cell cytotoxicity following recognition of MHC class I molecules on target cells. Different two-domain KIRs (KIR2Ds) recognise distinct subsets of HLA-C alleles. SPR analysis showed that, like other cell-cell recognition molecules interactions, the KIR2DL3 binds peptide-HLA-Cw7 with a low affinity ($Kd\sim10^{-5}M$), fast kinetics, and favourable entropic changes. In contrast, recent studies have shown that TCR/peptide-MHC interactions are characterised by slow kinetics and highly unfavourable entropic changes. Thus, although the TCR and KIRs both show allele- and peptide-specific MHC recognition, they bind with very different thermodynamic and kinetic properties.

Fcγ receptors (FcγR) are expressed on immunologically active cells, bind the Fc portion of IgG and contribute to phagocytosis, cytotoxicity and the clearance of immune complexes. SPR analysis showed that the human low-affinity FcγRs (FcγRIIa, FcγRIIb and FcγRIII) bind Fc with fast kinetics and a low affinity ($Kd\sim10^{-6}M$), as observed with other cell-cell recognition interactions, including KIR/HLA interactions. Interestingly, whereas the FcγRIIa/Fc and FcγRIIb/Fc interactions exhibited favourable entropic changes, comparable to the KIR/HLA interaction, the FcγRIII/Fc interaction was characterized by large unfavourable entropic changes.

Key words. Killer cell Ig-like receptor, Fcγ receptor, T cell receptor, MHC class I, Cell surface receptor, Molecular recognition, Surface plasmon resonance

Introduction

There are a wide variety of immunoreceptors possessing immunoreceptor tyrosine-based inhibitory motifs (ITIMs), which form what has been termed (Lanier 1998) the Inhibitory-receptor superfamily (IRS). Molecules in the IRS fall into two groups, (1) immunoglobulin(Ig)-like receptors, and (2) C-type lectin-like receptors. Many of these IRS molecules have activatory counterparts that have almost the same extracellular regions, but different intracellular regions. These typically contain one or more immunoreceptor tyrosine-based activatory motifs (ITAMs) or are associated with molecules possessing ITAMs.

Ig-like members of the IRS whose crystal structures are known include human killer cell Ig-like receptors (KIRs) [KIR2DL1 (Fan et al. 1997), 2DL2 (Snyder et al. 1999), 2DL3 (Maenaka et al. 1999b)] and the FcγRIIb (Sondermann et al. 1999). Additionally the structures of the activating FcRs, FcγRIIa (Maxwell et al. 1999), FcγRIII (Sondermann et al. 2000) and FcεRIα (Garman et al. 1998), have been determined. The Ig domains in these receptors have a characteristic and unique structure, intermediate between the I-set and C2-set Ig folds (Fig. 1b). Although sequence similarity among these Ig-like receptors is not high, this structural similarity suggests they have the same common ancestor receptor. The recently solved structures of the complexes KIR2DL2/HLA-Cw3 (Boyington et al. 2000) and FcγRIII/Fc (Sondermann et al. 2000) reveal that they bind in a different manner to their ligands (Fig. 1a). While the KIR molecule uses both Ig like domains for binding MHC class I, the FcγR uses only the membrane-proximal domain for binding the Fc.

Killer cell immunoglobulin-like receptors (KIRs, killer cell inhibitory receptors)

NK cells and cytotoxic T lymphocytes (CTLs) have complementary roles in the cellular immune response. CTLs kill cells presenting non-self peptides on MHC class I molecules whereas NK cells kill cells deficient in MHC class I molecules (Lanier 1998; Long 1999). In humans a family of killer cell Ig receptors (KIRs) have been identified on NK cells, and a subset of T cells, that check the expression of MHC class I molecules on the target cells and regulate (both activate and inhibit) the cellular function. The KIR gene family is clustered on human chromosome 19q13.4 and the number of genes (~10) varies between individuals. These characteristics, and the absence of homologous genes in rodents, suggest that the KIR family evolved recently, probably driven by the rapid evolution of MHC class I molecules.

Members of KIR family have either two (KIR2D) or three (KIR3D) Ig-like extracellular domains. They can be further grouped into (1) inhibitory receptors (KIR2DL or KIR3DL) with long cytoplasmic tails including immunoreceptor tyrosine-based inhibitory motifs (ITIM) that transduce inhibitory signals (Lanier, 1998) and (2) activating receptors (KIR2DS or KIR3DS) with short cytoplasmic tails that mediate association with DAP12, which contains immunoreceptor tyrosine-based activation motifs (ITAM) that transduce activatory signals (Lanier et al. 1998).

While the KIR2D receptors bind HLA-C alleles, KIR3D receptors bind HLA-A and B alleles (Lanier et al. 1998). The KIR2D binding site on HLA-C, which has been delineated by functional studies and crystallographic analysis of the KIR2DL2/HLA-Cw3 complex structure, includes residues around the C-terminal half of the $\alpha 1$ helix and the peptide binding site. KIR2DL1 binds to 'group 1' HLA-C alleles (Cw2, Cw4, Cw5, Cw6), which have a Lys at position 80, whereas KIR2DL2/3 bind to 'group 2' HLA-C alleles (Cw1, Cw3, Cw7, Cw8), which have an Asn at position 80. KIR recognition of peptide-MHC class I has been shown to depend to some extent on the peptide presented on the MHC molecule.

Binding characteristics of KIR/MHC interactions

The soluble forms of KIR2Ds and MHC class I molecules, which are produced in E.coli as inclusion bodies and refolded, were used for surface plasmon resonance (SPR) analysis (Maenaka et al. 1999a). Consistent with several cellular studies that have demonstrated that NK cell recognition is dependent on the peptide as well as the MHC on target cells, we found that the affinity of KIR2DL3 binding to HLA-Cw7-peptide was considerably affected by the nature of the peptide (Table 1). KIR2DL3 bound to HLA-Cw7- NKADVILKY, -RYRPGTVAL, and - KYFDEHYEY with a Kd ~ 7 μM, ~ 115 μM, and >3000 μM, respectively (Table 1). Taken together with the report by Mandelboim et al (Mandelboim et al. 1997) showing that killing by several NK clones specific for group 2 HLA-C alleles was inhibited by HLA-Cw7 loaded with the RYRPGTVAL peptide, this suggests that affinities as low as Kd ~ 115 μM are enough to mediate inhibition. Boyington et al. (2000) showed that the amino acid at the position 8 of the bound peptide affected this binding (see Table 1), consistent with the crystal structure of KIR2DL2/HLA-Cw3 complex which clearly showed that the loops of KIR2DL2 directly interact with the position 8. We also showed that, although KIR2DL molecules are grouped according to whether they bind to group 1 (KIR2DL1) or group 2 (KIR2DL2/3) HLA-C alleles, KIR2DL1 can bind to group 2 HLA-Cw7, albeit with ~10 fold lower affinity than KIR2DL3.

Several lines of evidence indicate that the stoichiometry of KIR2D/MHC class I interactions is likely to be 1:1: (1) the KIR2DL3/HLA-Cw7 complex migrated on gel filtration at the position expected for a 1:1 complex; (2) the standard Langmuir 1:1 binding model fits very well to the equilibrium binding and kinetic data; and (3) the crystal structure of the KIR2DL2/HLA-Cw3 showed 1:1 complex (Boyington et al. 2000).

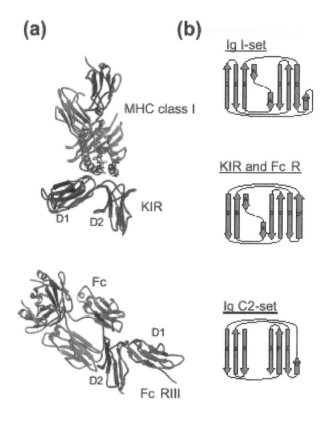

Fig. 1. a Crystal structures of KIR2DL2/HLA-Cw3 (top) and FcγRIII/Fc (bottom) complexes. These structures are superimposed on the basis of the D2 (domain proximal to the membrane) of each receptor. This figure are produced by Bobscript (Esnouf, 1997) and Raster3D (Merritt and Murphy, 1994). **b** Structural topology of Ig-like domains, I-set (top), C2-set (bottom) and KIR-FcγR type (middle). Each β strand is indicated by an arrow.

Table 1. Summary of affinity of KIR2D/MHC class I interactions

Receptor	MHC class I	Kd (μM)	References
KIR2DL3	HLA-Cw7-KYFDEHYEY	>3000	[1,2]
	-NKADVILKY	~110	[1]
	-RYRPGTVAL	~7	[1,2]
KIR2DL1	HLA-Cw7-RYRPGTVAL	~120	[1]
	HLA-Cw6-YQFTGIKKY	~10	[2]
KIR2DL2	HLA-Cw3-GAVDPLLAL	9.5	[3]
	-GAVDPLLSL	42	[3]
	-GAVDPLLVL	530	[3]
	-GAVDPLLYL	>600	[3]
	-AAADAAAAL	49	[3]
	-TAMDVVYAL	38	[3]
	-QAISPRTL	74	[3]

[1] (Maenaka et al. 1999a) [2] (Vales-Gomez et al. 1998) [3] (Boyington et al. 2000)

Table 2. Summary of affinity and kinetic data for lymphocyte cell-cell recognition molecules

Interaction	k_{on} ($M^{-1}s^{-1}$)	k_{off} (s^{-1})	Kd (μM)	References
KIR2D/HLA-C	~200 000	1.2-2	5-10	[1,2,3]
sFcγR/hFc1	~400 000	0.31-1.0	0.7-2.5	[4]
TCR/peptide-MHC	900-20 000	0.01-0.1	1-90	[5,6]
CD8αα/MHC class I	>100 000	>18	~200	[7]
sFcαR/Fc	~26 800	0.028	1.0	[8]
CD2/CD58	>400 000	>4	~10	[9]
Mouse CD48/CD2	>120 000	>11	~90	[10]
Mouse CD48/2B4	>200 000	>3	~15	[11]
CD28/CD80	>660 000	>1.6	~4	[12]
CTLA-4/CD80	>940 000	>0.4	0.46	[12]
CD62L/GlyCAM-1	>100 000	>10	108	[13]
ICAM-1/LFA-1	200 000	0.1	0.5	[14]

[1](Maenaka et al. 1999a), [2](Vales-Gomez et al. 1998), [3](Boyington et al. 2000), [4](Maenaka et al. unpublished), [5](Davis et al. 1998a), [6](Willcox et al. 1999), [7](Wyer et al. 1999), [8](Wines et al. 1999), [9](van der Merwe et al. 1994), [10](Davis et al. 1998b), [11](Brown et al. 1998), [12](van der Merwe et al. 1997), [13](Nicholson et al. 1998), [14](Tominaga et al. 1998)

The interactions between soluble forms of KIR2Ds and the peptide-MHC complex showed a low affinity (Kd ~ 10 μM). These affinities are similar to those of many other cell-cell recognition molecules, including TCR/peptide-MHC interactions (Table 2). However, while the low affinity of TCR/peptide-MHC interactions is a consequence of a relatively slow k_{on} rather than a fast k_{off} (Davis et al. 1998a; Willcox et al. 1999), the low affinity of KIR2D/peptide-MHC interactions is a consequence of a faster k_{off}, being typical of other cell surface protein/protein interactions (Table 2). Furthermore thermodynamic parameters of the KIR/peptide-MHC and TCR/peptide-MHC interactions were very different (Maenaka *et al.*, 1999a). While the TCR binding showed the large, unfavourable entropic changes compensated for by even larger favourable enthalpic changes, KIR binding showed both favourable entropic and enthalpic changes, which are typical of protein/protein interactions including low-affinity interactions between cell-cell recognition molecules, such as the CD2/CD48 interaction (J. Ladbury, P.A. van der Merwe, and S.J. Davis, unpublished data). These results suggest that KIR/peptide-MHC binding is likely to be a rigid-body association or an association which maintains the flexibility of each component of the complex.

Fcγ receptors

Fcγ receptors (FcγR) are membrane glycoproteins expressed on all immunologically active cells and bind the Fc portion of the main serum immunoglobulin, IgG, to trigger a wide range of immune response. In vitro experiments have shown that these soluble receptor forms can be used as a tool for prognosis and disease stage of systemic lupus erythematodes (SLE) (Hutin et al. 1994), rheumatoid arthritis (RA) (Fleit et al. 1992), multiple myeloma (MM) (Mathiot et al. 1993), and HIV infection (Khayat et al. 1990).

FcγRs belong to the immunoglobulin (Ig) superfamily and have been classified into three classes: FcγRI (CD64), also known as the high affinity receptor (Kd ~10^{-8}M), contains three Ig-like domains in the extracellular region while FcγRII (CD32) and FcγRIII (CD16), the low affinity receptors (Kd ~10^{-6}M), both contain two Ig-like domains. FcγRII is expressed on nearly all immunocompetent cells while the transmembrane form of FcγRIII (FcγRIIIa) is expressed on natural killer (NK) cells, macrophages and mast cells. These receptors bind aggregated IgG on the surface of target cells thereby enhancing phagocytosis and antibody dependent cellular cytotoxicity (ADCC).

Binding characteristics of low affinity FcγRs/Fc interactions

The ectodomains of FcγRs were expressed in E.coli and refolded in vitro. Binding to IgG or Fc was analysed by SPR. The affinity of sFcγRs binding to immobilized Fc was measured by equilibrium binding analysis. All FcγRs showed similar affinities ($K_D = 0.7$-$2.5\mu M$) with very fast dissociation rates (Table 2) (Maenaka et al. unpublished).

We estimated the enthalpy change (ΔH) which accompanies FcγRs binding to Fc by van't Hoff analysis, which involves measuring the dependence of affinity on temperature. For FcγRIIa and IIb, at 25°C favourable enthalpic and entropic changes contribute to the binding energy. In contrast FcγRIII binding to IgG or Fc was characterised by large unfavourable entropic changes. This was compensated for by even larger favourable enthalpic changes thus yielding a similar binding energy to FcγRII/IgG interactions. Thus although the binding affinity and kinetics are similar in the FcγRs studied, the thermodynamic characteristics are quite different (Maenaka et al. unpublished).

The standard 1:1 Langmuir binding model fits very well to the equilibrium binding and kinetic data of SPR analysis. This is consistent with the results of other studies indicating that one FcγR binds one Fc (Kato et al. 2000) as well as the recent crystal structure of FcγRIII/Fc complex (Sondermann et al. 2000). These findings suggest that, although, as a dimer, IgG might be expected to contain two binding sites for a FcγR, induced conformational changes in the second potential binding site could prevent a second FcγR from binding to a FcγR/IgG complex.

Conclusion and perspective for Ig-like receptors

This review focused on two groups of Ig-like receptors within the IRS that share a unique structural topology (Fig. 1). The interactions of KIRs and FcγRs with their natural ligands showed binding characteristics typical of cell surface receptors, and distinct from TCR/pep-MHC interactions (Table 2). Although recognizing overlapping regions of peptide-MHC class I molecules, TCR binding appears to be accompanied by a reduction in conformational flexibility at the binding interface (Garcia et al. 1998; Willcox et al. 1999) whereas the KIR binding shows no evidence for such a reduction (Maenaka et al. 1999a). It has been proposed that this difference arises from the fact that, unlike KIRs, the genes encoding the highly variable binding loops of TCRs are produced in a semi-random manner during T cell development (Willcox et al. 1999). TCRs thus do not have the opportunity available to germ-line encoded receptors such as KIRs to evolve structural stability.

The recent crystal structure of the KIR2DL2/HLA-Cw3 complex (Boyington et al. 2000) revealed a predominance of the electrostatic interactions at the interface, which resembles other cell surface receptor complexes such as CD2/CD58 (Wang et al. 1999). However, the FcγRIII/Fc complex structure does not show a predominance of electrostatic interactions, indicating this is not a universal feature of low affinity cell-surface interactions.

In addition to their structural similarity, Wines and co-workers (Wines et al. 1999) have pointed out that KIRs and Fc receptors (e.g. FcγRII, FcγRIII, FcεRIα, FcαR) use the FG loop for the ligand binding (Fig. 1a). While sharing a unique structural topology and using the same FG loop for ligand binding, there are differences in their domain arrangements and the ligand binding sites (Fig. 1a), suggesting that KIRs and FcγRs may have a related but distinct ligand binding mechanism.

Finally, in view of their gene localization, Kasahara has suggested that human chromosome 19q13.4, which includes the KIR/ILT family genes, may be paralogous to 1q23, where the FcγRs genes are located (Kasahara 1999).

Acknowledgements

We thank I. Kumagai, M. Matsushima, T. Juji, K. Tadokoro, K. Tokunaga, T. Yabe, K. Tsumoto, T. Nakayama, K. Harlos and M. Kasahara for their advice and discussion. K.M. is supported by a grant-in-aid for scientific research on Priority Areas (C) Genome Information Science from the Ministry of Education, Science, Sports and Culture of Japan. D.I.S. and E.Y.J. are supported by MRC and Royal Society respectively.

References

Boyington JC, Motyka SA, Schuck P, Brooks AG, Sun PD (2000) Crystal structure of an NK cell immunoglobulin-like receptor in complex with its class I MHC ligand. Nature 405:537-543

Brown MH, Boles K, van der Merwe PA, Kumar V, Mathew PA, Barclay AN (1998) 2B4, the natural killer and T cell immunoglobulin superfamily surface protein, is a ligand for CD48. J Exp Med 188:2083-2090

Davis MM, Boniface JJ, Reich Z, Lyons D, Hampl J, Arden B, Chien Y (1998a) Ligand recognition by alpha beta T cell receptors. Annu Rev Immunol 16:523-544

Davis SJ, Ikemizu S, Wild MK, van der Merwe PA (1998b) CD2 and the nature of protein interactions mediating cell-cell recognition. Immunol Rev 163:217-236

Esnouf RM (1997) An extensively modified version of MolScript that includes greatly enhanced coloring capabilities. J Mol Graph Model 15:132-134

Fan QR, Mosyak L, Winter CC, Wagtmann N, Long EO, Wiley DC (1997) Structure of the inhibitory receptor for human natural killer cells resembles haematopoietic receptors. Nature 389:96-100

Fleit HB, Kobasiuk CD, Daly C, Furie R, Levy PC, Webster RO (1992) A soluble form of Fc γ RIII is present in human serum and other body fluids and is elevated at sites of inflammation. Blood 79:2721-2728

Garcia KC, Degano M, Pease LR, Huang M, Peterson PA, Teyton L, Wilson IA (1998) Structural basis of plasticity in T cell receptor recognition of a self peptide-MHC antigen. Science 279:1166-1172

Garman SC, Kinet JP, Jardetzky TS (1998) Crystal structure of the human high-affinity IgE receptor. Cell 95:951-961

Hutin P, Lamour A, Pennec YL, Soubrane C, Dien G, Khayat D, Youinou P (1994) Cell-free Fcγ receptor III in sera from patients with systemic lupus erythematosus: correlation with clinical and biological features. Int Arch Allergy Immunol 103:23-27

Kasahara M (1999) The chromosomal duplication model of the major histocompatibility complex. Immunol Rev 167:17-32

Kato K, Sautes-Fridman C, Yamada W, Kobayashi K, Uchiyama S, Kim H, Enokizono J, Galinha A, Kobayashi Y, Fridman WH, Arata Y, Shimada I (2000) Structural basis of the interaction between IgG and Fcγ receptors. J Mol Biol 295:213-224

Khayat D, Soubrane C, Andrieu JM, Visonneau S, Eme D, Tourani JM, Beldjord K, Weil M, Fernandez E, Jacquillat C (1990) Changes of soluble CD16 levels in serum of HIV-infected patients: correlation with clinical and biologic prognostic factors. J Infect Dis 161:430-435

Lanier LL (1998) NK cell receptors. Annu Rev Immunol 16:359-393

Lanier LL, Corliss BC, Wu J, Leong C, Phillips JH (1998) Immunoreceptor DAP12 bearing a tyrosine-based activation motif is involved in activating NK cells. Nature 391:703-707

Long EO (1999) Regulation of immune responses through inhibitory receptors. Annu Rev Immunol 17:875-904

Maenaka K, Juji T, Nakayama T, Wyer JR, Gao GF, Maenaka T, Zaccai NR, Kikuchi A, Yabe T, Tokunaga K, Tadokoro K, Stuart DI, Jones EY, van der Merwe PA (1999a) Killer cell immunoglobulin receptors and T cell receptors bind peptide-major histocompatibility complex class I with distinct thermodynamic and kinetic properties. J Biol Chem 274:28329-28334

Maenaka K, Juji T, Stuart DI, Jones EY (1999b) Crystal structure of the human p58 killer cell inhibitory receptor (KIR2DL3) specific for HLA-Cw3-related MHC class I. Structure 7:391-398

Mandelboim O, Wilson SB, Vales-Gomez M, Reyburn HT, Strominger JL (1997) Self and viral peptides can initiate lysis by autologous natural killer cells. Proc Natl Acad Sci USA 94:4604-4609

Mathiot C, Teillaud JL, Elmalek M, Mosseri V, Euller-Ziegler L, Daragon A, Grosbois B, Michaux JL, Facon T, Bernard JF, et al. (1993) Correlation between soluble serum CD16 (sCD16) levels and disease stage in patients with multiple myeloma. J Clin Immunol 13:41-48

Maxwell KF, Powell MS, Hulett MD, Barton PA, McKenzie IF, Garrett TP, Hogarth PM (1999) Crystal structure of the human leukocyte Fc receptor, Fc gammaRIIa. Nat Struct Biol 6:437-442

Merritt EA, Murphy MEP (1994) Raster3D version-2.0 - A program for photorealistic molecular graphics. Acta Cryst D50:869-873

Nicholson MW, Barclay AN, Singer MS, Rosen SD, van der Merwe PA (1998) Affinity and kinetic analysis of L-selectin (CD62L) binding to GlyCAM-1. J Biol Chem 273:763-770

Snyder GA, Brooks AG, Sun PD (1999) Crystal structure of the HLA-Cw3 allotype-specific killer cell inhibitory receptor KIR2DL2. Proc Natl Acad Sci USA 96:3864-3869

Sondermann P, Huber R, Oosthuizen V, Jacob U (2000) The 3.2-A crystal structure of the human IgG1 Fc fragment-FcγRIII. Nature 406:267-273

Sondermann P, Huber R, Jacob U (1999) Crystal structure of the soluble form of the human fcgamma-receptor IIb: a new member of the immunoglobulin superfamily at 1.7 A resolution. Embo J 18:1095-1103

Tominaga Y, Kita Y, Satoh A, Asai S, Kato K, Ishikawa K, Horiuchi T, Takashi T (1998) Affinity and kinetic analysis of the molecular interaction of ICAM-1 and leukocyte function-associated antigen-1. J Immunol 161:4016-4022

Vales-Gomez M, Reyburn HT, Mandelboim M, Strominger JL (1998) Kinetics of interaction of HLA-C ligands with natural killer cell inhibitory receptors. Immunity 9:337-344

van der Merwe PA, Barclay AN, Mason DW, Davies EA, Morgan BP, Tone M, Krishnam AKC, Ianelli C, Davis SJ (1994) The human cell-adhesion molecule CD2 binds CD58 with a very low affinity and an extremely fast dissociation rate but does not bind CD48 or CD59. Biochemistry 33:10149-10160

van der Merwe PA, Bodian DL, Daenke S, Linsley P, Davis SJ (1997) CD80 (B7-1) binds both CD28 and CTLA-4 with a low affinity and very fast kinetics. J Exp Med 185:393-403

Wang JH, Smolyar A, Tan K, Liu JH, Kim M, Sun ZY, Wagner G, Reinherz EL (1999) Structure of a heterophilic adhesion complex between the human CD2 and CD58 (LFA-3) counterreceptors. Cell 97:791-803

Willcox BE, Gao GF, Wyer JR, Ladbury JE, Bell JI, Jakobsen BK, van der Merwe PA (1999) TCR binding to peptide-MHC stabilizes a flexible recognition interface. Immunity 10:357-365

Wines BD, Hulett MD, Jamieson GP, Trist HM, Spratt JM, Hogarth PM (1999) Identification of residues in the first domain of human Fc α receptor essential for interaction with IgA. J Immunol 162:2146-2153

Wyer JR, Willcox BE, Gao GF, Gerth UC, Davis SJ, Bell JI, van der Merwe PA, Jakobsen BK (1999) T cell receptor and coreceptor CD8 αα bind peptide-MHC independently and with distinct kinetics. Immunity 10:219-225

7
Ig modules as discrete structural units to exploit functional and structural aspects of Ig-like receptors

Luca Vangelista and Oscar Burrone

International Centre for Genetic Engineering and Biotechnology (ICGEB), Molecular Immunology Group, Padriciano 99, 34012-Trieste, Italy

Summary. Ig-like receptors derive their name from the presence of immunoglobulin-like domains in their extracellular moieties. The Ig module is the most widespread protein fold in nature, with more than a thousand members found in the most diverse and evolutionarily distant protein families. The simple architecture of the Ig fold makes it ideal for structural and functional studies as isolated module. In the field of Fc receptors, we describe here two examples of single Ig modules usage. The first application helps the understanding of FcεRI binding and folding by studying the membrane-proximal Ig module (α2) of the receptor α-chain. In the second strategy, the heavy chain C-terminal Ig module of human immunoglobulins (eg, CH3 for IgG and CH4 for IgE) allows the design of dimeric molecules with good potentials for clinical and biotechnological applications.

Key words. Ig module, Fc, Fold, Binding, Dimerisation

The Ig module

Immunoglobulin (Ig) domains can be defined as protein modules since they are found in a large variety of proteins and fold into a discrete unit defined within their N- and C-terminus (reviewed in Doolittle and Bork 1993). In fact, the Ig module is the most widespread protein fold in nature, with more than a thousand members found in the most diverse and evolutionarily distant protein families. This fold consists of two β-sheets forming a characteristic sandwich with a core structure (Fig. 1b) common to all members of the Ig superfamily (IgSF) (Bork et al. 1994). According to their β strands topology, IgSF members have been classified into distinct sets (Williams and Barclay 1988; Harpaz and Chothia

1994). Overall, the Ig module has proven to be an ideal scaffold for many protein functions, ranging from protein-protein interactions (eg, antibody interactions) to muscle elasticity (titin I-band Ig domains). Presence of Ig modules in distantly related organisms has been attributed to both convergent evolution (De Marino et al. 1999) and divergence from a common ancestor (Bateman 1996), according to biocomputing analysis of each single case. Convergent evolution and extensive usage are both the likely result of thermodynamic parameters favouring the Ig fold.

The simple architecture of the Ig fold makes it ideal for structural and functional studies as isolated module, a concept corroborated by numerous reports. The possibility to dissect and analyse single Ig blocks from complex proteins often accelerates and simplifies enormously their structural and functional studies (reviewed in Campbell and Downing 1994). Protein modular dissection has been successfully performed on many different proteins and modules type. Moreover, the analysis of single Ig modules allowed detailed investigation on their folding pathways (Clarke et al. 1999).

There are members of the IgSF naturally occurring as proteins made by a single Ig-like module. This is the case for two different allergen families of which Der f 2 (a dust mite allergen) and Phl p 2 (a pollen allergen) 3D structures have been recently determined (Ichikawa et al. 1998; De Marino et al. 1999). β_2-microglobulin (Becker and Reeke 1985) and the major nematode sperm protein (Bullock et al. 1996) are also soluble single Ig modules, although they exist in complexes *via* non-covalent interactions. Finally, Thy-1 (Campbell et al. 1979), the very recently cloned Fc receptor for IgA and IgM (Fcα/μ) (Shibuya et al. 2000; Kinet and Launay 2000) and CTLA4 (Metzler et al. 1997; Ostrov et al. 2000) are Ig-like receptors presenting a single Ig module as their extracellular moiety.

In the field of Fc receptors, we describe here two examples of single Ig module analysis and usage.

The membrane-proximal Ig module of FcϵRI α-chain

FcϵRI, the high affinity receptor for IgE, is a multichain membrane receptor (existing as $\alpha\beta\gamma_2$ or $\alpha\gamma_2$) with two Ig modules in the extracellular portion of its α-chain, the IgE binding moiety (Fig. 2). Efforts to characterise the IgE binding mode of human FcϵRI evidenced the binding site to be located within the membrane-proximal α-chain Ig module (reviewed in Sutton and Gould 1993). Recently, the structure of human IgE-FcϵRIα complex has been solved (Garman et al. 2000), representing a breakthrough for the understanding of this interaction. IgE appears now to interact with two distinct binding sites on FcϵRIα membrane-proximal Ig domain.

We expressed isolated FcεRIα membrane-proximal Ig module (named here α2) as a strategy to ascertain and compare α2 binding properties with the α1α2 module pair (Vangelista et al. 1999).

FcεRI is a key player in IgE homeostasis and the pathology of atopic allergy (reviewed in Kinet 1999). Therefore, the dissection of this multichain membrane-bound receptor for the construction of simpler soluble versions (Fig. 2) could potentially lead to anti-allergic protein drugs. Isolated α2 ultimately represents a powerful candidate for manifold studies, ranging from structural characterisation to binding competition with the native receptor. In a medical context, soluble α2 could act as an IgE scavenger, blocking or preventing the in vivo cell receptor occupancy. Yet, the in vivo half life of this molecule would likely be too short for a proper efficacy. In this context, an alternative strategy, discussed in the second paragraph, could account for a possible solution of this problem.

Many efforts have been made to obtain α2 from a prokaryotic source (Vangelista 1999), but misfolding and aggregation of the unglycosylated protein confirmed the reported importance of glycosylation for folding and secretion of a fully active receptor (Letourneur et al. 1995; Albrecht et al. 2000). An eukaryotic version of α2 was obtained as soluble glycosylated protein both from insect cells (Vangelista et al. 1999) and mammalian cells (Vangelista et al. in preparation).

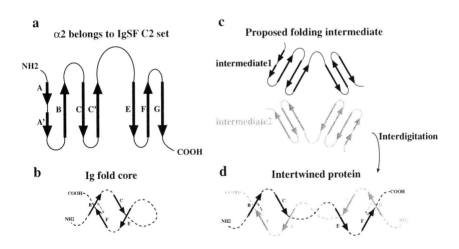

Fig. 1a-d. Folding of FcεRIα membrane-proximal Ig module. **a** According to its β-strands topology, α2 belongs to the IgSF C2 set. **b** The common IgSF core formed by the conserved BCEF strands (Bork et al. 1994) (S-S bond is indicated for α2 canonical cysteines). **c** Folding intermediates as proposed in Murray et al. (1995). **d** Intertwining of the two intermediates generates a dimer with the Ig fold (only Ig cores are shown).

Analysis of the protein under non-reducing conditions revealed an unexpected dimerisation *via* intermolecular disulphide bonds. Dimerisation occurs intracellularly and represents ~90% of the secreted protein. The monomeric fraction shows limited IgE binding, while the dimer does not bind IgE (Vangelista et al. in preparation). Both α2 monomer and dimer are terminally glycosylated and actively secreted, suggesting a degree of folding compatible with quality control mechanisms operating within the ER/Golgi compartments (Albrecht et al. 2000).

We thus predicted an alternative folding pathway leading to an intertwined α2 dimer where the cysteine residues, normally forming the canonical intramolecular Ig fold disulphide bond, react to yield a disulphide-bonded dimer (Fig. 1c,d).

Extended domain swapping leading to complete interdigitation of two monomers was reported for the N-terminal rat CD2 Ig module whose dimeric 3D structure was solved, showing the first β-sheet of one monomer assembled with the complementing one of the second monomer and *viceversa* (Murray et al. 1995). There are two major differences between the CD2 and the FcεRI case: i) the CD2 module was expressed in prokaryotic cells, while α2 was secreted by mammalian cells; ii) the CD2 module does not have the canonical Ig fold cysteine residues, while α2 does and forms the intermolecular bonds expected in an intertwined dimer.

In conclusion, the occurrence of an unexpected dimer of FcεRI α2 raised many questions on the folding and stability of this Ig domain. α2 dimerisation indicates an alternative folding for this protein, probably deriving from destabilisation of the monomeric form in the context of the single module. Since the α1α2 pair folds as a monomer, it can be assumed that the N-terminal Ig module of FcεRI α-chain (α1) could assist the folding of α2 in *cis*. It will be interesting to see whether α1 will rescue monomeric α2 also in *trans* when coexpressing the two modules as isolated entities. We therefore wish to study the effect of α1 proximity on α2 folding by coexpressing the two modules as single units (α1 in *trans* to α2). These studies should confirm the hypothesis that α1 acts as an intramolecular N-terminal chaperone-like block towards α2, as reported for other proteins (Ma et al. 2000).

The immunoglobulin heavy chain C-terminal module

In all secretory immunoglobulin isotypes, the heavy chain C-terminal Ig domain forms homo-dimers *via* non-covalent interactions. In the case of IgG, this Ig module (γCH3) has been characterised for its folding and dimerisation (Thies et al. 1999).

We made use of this molecular feature for the engineering of small immune proteins (SIPs) (Fig. 2) containing the antibody VL-VH antigen binding regions (in a single chain Fv format, scFv) fused to γCH3 (γ-SIPs) (Li et al. 1997). SIPs were shown to dimerise and retain full antibody binding capacity (Li et al. 1997). A similar version of chimeric antibody-like molecules was reported independently by another group (Hu et al. 1996; Wu et al. 2000), in which the IgG hinge region was interposed between the scFv and γCH3. Recent developments of our system, including SIPs (ε-SIPs) containing the dimerising C-terminal domain (εCH4) from the two secretory IgE isoforms, S1 and S2 (Batista et al. 1996), provided evidence of excellent characteristics for in vivo cancer diagnostics (Borsi et al. in preparation). SIPs containing εCH4 S2 are disulphide-bonded dimers, thanks to the C-terminal cysteine residue of IgE S2 isoform (Batista et al. 1996; Sepulveda and Burrone in preparation). Moreover, γ-SIPs have successfully been used in DNA vaccination protocols to induce protective anti-idiotypic antibody responses for immunotherapy of B-cell lymphomas (Benvenuti et al. 2000).

We then extended the engineering strategy to obtain membrane versions of these small antibody-like proteins (membrane SIPs). Membrane SIPs have, in addition to the dimerizing domain, the extracellular membrane-proximal domain (EMPD, the portion linking the Ig C-terminal domain to the transmembrane region) and the transmembrane and cytoplasmic regions of the original immunoglobulin (membrane IgG, IgE and IgA), thus targeting the proteins to the cell surface (Fig. 2). These constructs allowed the biochemical characterisation of the cysteines present in the EMPDs. As membrane SIPs do not have any intermolecular disulphide-forming cysteines, other than those in the EMPDs, we found and mapped intermolecular S-S bonds occurring within this stalk region in human IgE, IgA and IgG (Bestagno et al. submitted). In addition, membrane SIPs

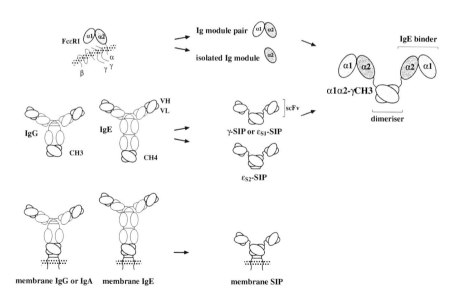

Fig. 2. Schematic representation of recombinant proteins generated by Ig modules usage.

60

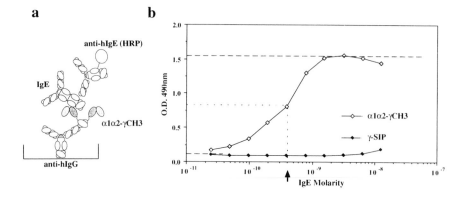

Fig. 3a,b. IgE binding by α1α2-γCH3. **a** Assay scheme. **b** Serial dilutions of human IgE allowed an estimation of α1α2-γCH3 binding affinity (arrow indicates 50% binding).

have been used to display different VL-VH idiotypes on the cell surface, providing a tool to analyse anti-idiotypic immune responses obtained by the SIP-DNA vaccination mentioned above (Benvenuti and Burrone submitted).

The final and most intriguing application of these versatile dimerising Ig domains consists in their use as dimerisers of protein ligands of biotechnological interest.

Our most immediate task was the fusion of human γCH3 with human FcεRI α1α2 module pair (Fig. 2) (Vangelista et al. in preparation). This protein (named α1α2-γCH3): i) is fully and terminally glycosylated by mammalian cells, ii) is actively secreted as a non-covalent dimer, iii) binds IgE with affinity comparable to the native receptor (Fig. 3b), iv) is fully human, and v) does not bind to Fcγ receptors. In fact, Fcγ receptors bind the γCH2 domain and part of the hinge region, as widely reported and finely demonstrated by the 3D structure of IgG-FcγRIII complex (Sondermann et al. 2000). Oppositely, binding to FcRn, which increases IgG in vivo half life, seems unlikely for SIPs and α1α2-γCH3, since the receptor binding site resides largely in γCH2 (West and Bjorkman 2000). Nevertheless, tumor-specific SIPs showed: i) enhanced target localisation compared to the entire antibodies containing the same VL-VH pair, and ii) extended in vivo half life compared to the same VL-VH scFvs (Borsi et al. in preparation). Similarly, it is likely that most of the properties conferred by human γCH3 to scFvs would be transferred to the human FcεRI α1α2 moiety, increasing its potentials for in vivo applications.

References

Albrecht B, Woisetschlager M, Robertson MW (2000) Export of the High Affinity IgE Receptor From the Endoplasmic Reticulum Depends on a Glycosylation-Mediated Quality Control Mechanism. J Immunol 165:5686-5694

Bateman A, Eddy SR, Chothia C (1996) Members of the immunoglobulin superfamily in bacteria. Protein Sci 5:1939-1941

Batista FD, Efremov DG, Burrone OR (1996) Characterization of a second secreted IgE isoform and identification of an asymmetric pathway of IgE assembly. Proc Natl Acad Sci U S A 93:3399-3404

Becker JW, Reeke GN Jr (1985) Three-dimensional structure of beta 2-microglobulin. Proc Natl Acad Sci U S A 82:4225-4229

Benvenuti F, Burrone OR, Efremov DG (2000) Anti-idiotypic DNA vaccines for lymphoma immunotherapy require the presence of both variable region genes for tumor protection. Gene Ther 7:605-611

Bork P, Holm L, Sander C (1994) The immunoglobulin fold. Structural classification, sequence patterns and common core. J Mol Biol 242:309-320

Bullock TL, Roberts TM, Stewart M (1996) 2.5 Å resolution crystal structure of the motile major sperm protein (MSP) of Ascaris suum. J Mol Biol 263:284-296

Campbell DG, Williams AF, Bayley PM, Reid KB (1979) Structural similarities between Thy-1 antigen from rat brain and immunoglobulin. Nature 282:341-342

Campbell ID, Downing AK (1994) Building protein structure and function from modular units. Trends Biotechnol 12:168-172

Clarke J, Cota E, Fowler SB, Hamill SJ (1999) Folding studies of immunoglobulin-like beta-sandwich proteins suggest that they share a common folding pathway. Structure Fold Des 7:1145-1153

De Marino S, Morelli MA, Fraternali F, Tamborini E, Musco G, Vrtala S, Dolecek C, Arosio P, Valenta R, Pastore A (1999) An immunoglobulin-like fold in a major plant allergen: the solution structure of Phl p 2 from timothy grass pollen. Structure Fold Des 7:943-952

Doolittle RF, Bork P (1993) Evolutionarily mobile modules in proteins. Sci Am 269:50-56

Garman SC, Wurzburg BA, Tarchevskaya SS, Kinet JP, Jardetzky TS (2000) Structure of the Fc fragment of human IgE bound to its high-affinity receptor Fc epsilonRI alpha. Nature 406:259-266

Harpaz Y, Chothia C (1994) Many of the immunoglobulin superfamily domains in cell adhesion molecules and surface receptors belong to a new structural set which is close to that containing variable domains. J Mol Biol 238:528-539

Hu S, Shively L, Raubitschek A, Sherman M, Williams LE, Wong JY, Shively JE, Wu AM (1996) Minibody: A novel engineered anti-carcinoembryonic antigen antibody fragment (single-chain Fv-CH3) which exhibits rapid, high-level targeting of xenografts. Cancer Res 56:3055-3061

Ichikawa S, Hatanaka H, Yuuki T, Iwamoto N, Kojima S, Nishiyama C, Ogura K, Okumura Y, Inagaki F (1998) Solution structure of Der f 2, the major mite allergen for atopic diseases. J Biol Chem 273:356-360

Kinet JP (1999) The high-affinity IgE receptor (FcεRI): from physiology to pathology. Annu Rev Immunol 17:931-972

Kinet JP, Launay P (2000) Fcα/μR: single member or first born in the family? Nat Immunol 1:371-372

Letourneur O, Sechi S, Willette-Brown J, Robertson MW, Kinet JP (1995) Glycosylation of human truncated FcεRI α chain is necessary for efficient folding in the endoplasmic reticulum. J Biol Chem 270:8249-8256

Li E, Pedraza A, Bestagno M, Mancardi S, Sanchez R, Burrone O (1997) Mammalian cell expression of dimeric small immune proteins (SIP). Protein Eng 10:731-736

Ma B, Tsai CJ, Nussinov R (2000) Binding and folding: in search of intramolecular chaperone-like building block fragments. Protein Eng 13: 617-627

Metzler WJ, Bajorath J, Fenderson W, Shaw SY, Constantine KL, Naemura J, Leytze G, Peach RJ, Lavoie TB, Mueller L, Linsley PS (1997) Solution structure of human CTLA-4 and delineation of a CD80/CD86 binding site conserved in CD28. Nat Struct Biol 4:527-531

Murray AJ, Lewis SJ, Barclay AN, Brady RL (1995) One sequence, two folds: a metastable structure of CD2. Proc Natl Acad Sci U S A 92:7337-7341

Ostrov DA, Shi W, Schwartz JC, Almo SC, Nathenson SG (2000) Structure of murine CTLA-4 and its role in modulating T cell responsiveness. Science 290:816-819

Shibuya A, Sakamoto N, Shimizu Y, Shibuya K, Osawa M, Hiroyama T, Eyre HJ, Sutherland GR, Endo Y, Fujita T, Miyabayashi T, Sakano S, Tsuji T, Nakayama E, Phillips JH, Lanier LL, Nakauchi H (2000) Fc α/μ receptor mediates endocytosis of IgM-coated microbes. Nat Immunol 1:441-446

Sondermann P, Huber R, Oosthuizen V, Jacob U (2000) The 3.2-A crystal structure of the human IgG1 Fc fragment-FcγRIII complex. Nature 406:267-273

Sutton BJ, Gould HJ (1993) The human IgE network. Nature 366:421-428

Thies MJ, Mayer J, Augustine JG, Frederick CA, Lilie H, Buchner J (1999) Folding and association of the antibody domain CH3: prolyl isomerization preceeds dimerization. J Mol Biol 293:67-79

Vangelista L (1999) Molecular and structural characterisation of key elements of the Type I allergic reaction: allergens, IgE and the high affinity Fcε receptor. PhD Thesis. Heidelberg University, Heidelberg, Germany

Vangelista L, Laffer S, Turek R, Gronlund H, Sperr WR, Valent P, Pastore A, Valenta R (1999) The immunoglobulin-like modules Cepsilon3 and alpha2 are the minimal units necessary for human IgE-FcepsilonRI interaction. J Clin Invest 103:1571-1578

West AP Jr, Bjorkman PJ (2000) Crystal structure and immunoglobulin G binding properties of the human major histocompatibility complex-related Fc receptor. Biochemistry 39:9698-9708

Williams AF, Barclay AN (1988) The immunoglobulin superfamily--domains for cell surface recognition. Annu Rev Immunol 6:381-405

Wu AM, Yazaki PJ, Tsai Sw, Nguyen K, Anderson AL, McCarthy DW, Welch MJ, Shively JE, Williams LE, Raubitschek AA, Wong JY, Toyokuni T, Phelps ME, Gambhir SS (2000) High-resolution microPET imaging of carcinoembryonic antigen-positive xenografts by using a copper-64-labeled engineered antibody fragment. Proc Natl Acad Sci U S A 97:8495-8500

8
Structural studies on the leukocyte co-stimulatory molecule, B7-1

Shinji Ikemizu[1], E. Yvonne Jones[1,2], David I. Stuart[1,2] and Simon J. Davis[3]

[1]Division of Structural Biology, Wellcome Trust Centre for Human Genetics, The University of Oxford, Roosevelt Drive, Oxford OX3 7BN, United Kingdom
[2]Oxford Centre for Molecular Sciences, The University of Oxford, New Chemistry Laboratory, South Parks Road, Oxford, OX1 3QT, United Kingdom
[3]Nuffield Department of Clinical Medicine, The University of Oxford, John Radcliffe Hospital, Headington, Oxford OX3 9DU, United Kingdom

Summary. B7-1 and B7-2 are glycoproteins expressed on antigen presenting cells. The binding of these molecules to the T-cell homodimers, CD28 and CTLA-4, generate 'costimulatory' and inhibitory signals in T cells, respectively. The crystal structure of the extracellular region of B7-1 (sB7-1), solved to 3 Å resolution, consists of a novel combination of two Ig-like domains, one characteristic of adhesion molecules and the other previously seen only in antigen receptors. In the crystal lattice, sB7-1 unexpectedly forms parallel, 2-fold rotationally symmetric homodimers. The structural data suggest a mechanism whereby the avidity-enhanced binding of B7-1 and CTLA-4 homodimers, along with the relatively high affinity of these interactions, favours the formation of very stable inhibitory signaling complexes.

Key words. B7-1, CD80, Co-stimulatory molecule, Crystal structure

Introduction

B7-1 and B7-2 each consist of single V-like and C-like immunoglobulin superfamily (IgSF) domains. Their ligands, CD28 and CTLA-4, are also structurally related and are expressed at the cell surface as homodimers of single V-like IgSF domains. A third CD28-like molecule, ICOS, interacts with another B7-related molecule, but the analysis of transgenic mice indicates that B7-1 and B7-2 are the only functional ligands of CD28 and CTLA-4. The affinities of these interactions differ substantially. For example, human CTLA-4 binds B7-1 with a solution K_d of 0.2-0.4 μM (van der Merwe et al. 1997), whereas the affinity of

CD28 for B7-2 is 40-100 - fold lower [B7-1/CD28 and B7-2/CTLA-4 interactions each have intermediate affinities (K_d = 4 µM)].

The expression of B7-1, B7-2, CD28 and CTLA-4 is tightly regulated: whereas CD28 is constitutively expressed on resting human T cells and B7-2 is rapidly induced on antigen presenting cells early in immune responses, the expression of both B7-1 and CTLA-4 is considerably delayed (reviewed by Lenschow et al. 1996). Interactions of the B7 molecules with CD28 generate costimulatory signals amplifying T cell receptor (TCR) signaling and preventing anergy, whereas interactions with CTLA-4 induce powerful inhibitory signals in T cells. CD28-dependent costimulation is poorly understood but recent work implicates the bulk recruitment of cell surface molecules and kinase-rich rafts to the site of TCR engagement, favouring receptor phosphorylation and signaling. Conversely, CTLA-4 inhibits signal transduction by recruiting the tyrosine phosphatase SHP-2 to the TCR resulting in dephosphorylation of the ζ-chain of the complex and components of the RAS signaling pathway, and by interfering with distal events in the CD28 signaling pathway.

Whilst the opposing effects of CD28 and CTLA-4 are clear-cut, distinct functions for B7-1 and B7-2 have yet to be defined. A role in TH0, TH1 or TH2 differentiation has been proposed, but other work suggests that B7-1 and B7-2 determine the magnitude of costimulatory signals rather than the outcome of TH subset differentiation (reviewed by McAdam et al. 1998). Supporting this view, quantitative rather than qualitative effects on tyrosine phosphorylation follow the stimulation of T cells with artificial antigen-presenting cells expressing B7-1 or B7-2. Moreover, gene disruption studies reveal considerable overlap in the costimulatory functions of B7-1 and B7-2. A third possibility is that, rather than having distinct CD28-dependent costimulatory roles, the key functional differences concern the strength and/or mode of binding of B7-2 and B7-1 to CD28 and CTLA-4.

Protein expression and crystallization

A construct encoding a histidine-tagged form of soluble B7-1 (sB7-1), consisting of residues 1 to 201 of the native protein and bearing a deleted COOH-terminal glycosylation site (Asn-198), was expressed in Lec3.2.8.1 cells in the presence of 0.5 mM N-butyldeoxynojirimycin, as described (Butters et al. 1999). After deglycosylation with endoglycosidase H, sB7-1 bound CTLA-4 with wild-type affinity. Selenomethionine (SeMet) labelling was done according to the method of May et al. (1998). After removal of the histidine tag with carboxypeptidase A, data collection-quality wild-type and SeMet-labelled sB7-1 crystals grew readily in sitting drops containing 2 µl of protein (at 20-30 O.D. in 10 mM Hepes, 140 mM NaCl, 0.05% NaN_3, pH 7.4.) and 2 µl of reservoir solution (typically a 10%-90% v/v dilution of a stock solution containing 28% PEG 400, 0.1M Na Hepes pH 7.5, 0.2M $CaCl_2$).

Structure determination

Diffraction data to 3Å resolution were collected at 3 wavelengths (λ = 0.8855 Å, 0.97930 Å and 0.97955 Å) from one cryo-cooled SeMet-labelled sB7-1 crystal at station BM14 of the European Synchrotron Radiation Facility, Grenoble, France. The crystal belonged to space group I4$_1$22 and had cell dimensions a=b=57.3 Å, c=298.9 Å. The diffraction data were processed and scaled with the HKL program suite. The structure was solved to 3.0 Å resolution using multiple-wavelength anamolous dispersion (MAD) methods and the program SOLVE, then the initial phases were improved with programs DM and SOLOMON. The structure was built using program O. Refinement with CNS used a set of Fobs constructed by merging the diffraction data (including Bjivoet pairs) from all three MAD wavelengths. Refinement of the atomic positions and tightly restrained individual isotropic B-factors resulted in a final crystallographic R-factor and R-free of 23.7% and 28.2%, respectively.

The structure of sB7-1

Overall structure

The sB7-1 monomer is a slender molecule, with dimensions of 23 x 30 x 90Å^3, consisting of two anti-parallel β-sandwich IgSF domains joined by a short linker (Fig. 1A). Thr-199 forms the last mainchain hydrogen bond in the molecule and Ala-200 is well ordered but extends into solution. Between this residue and the probable first residue of the transmembrane domain (Leu-209), the native protein sequence consists of eight, mostly hydrophilic, residues likely to form an extended stalk linking the protein to the cell surface. Overall, the organisation and dimensions of sB7-1 resemble those of the extracellular region of CD2, although there is a significant change in relative domain orientation. The reduced tilt of domain 1 relative to domain 2 positions the ligand binding AGFCC'C" face of sB7-1 normal to the long axis of the molecule, rather than, as in CD2 parallel with it (Jones et al. 1992; Bodian et al. 1994). Such differences may be related to the sequence of expression of these genes. Maximum exposure of the ligand binding face at the "top" of the molecule may be crucial for establishing the initial contact zone between T cells and antigen presenting cells by CD2. B7-1 appears after formation of the contact zone, however, when the interacting cell surfaces are already optimally aligned for interactions involving molecules the size of B7-1, the T cell receptor and their respective ligands. An important consequence of the essentially parallel alignment of d1 and d2 of B7-1 is that it exposes the non-ligand binding DEB face of d1, allowing the formation of parallel, compact homodimers. The formation of equivalent dimers by CD2 is precluded by the 42°

tilt in the position of d1 relative to the long axis of the molecule (Jones et al. 1992; Bodian et al. 1994).

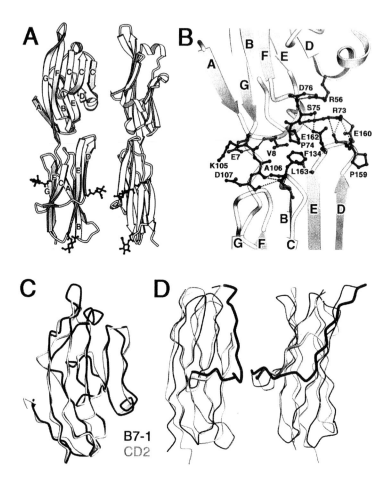

Fig. 1. The structure of sB7-1. (A) Two orthogonal views, shown in ribbon format, of the sB7-1 monomer. Asparagine residues that are part of glycosylation sequons are represented with modeled N-acetylglucosamine residues where electron density was sufficiently clear. (B) The inter-domain region of B7-1. The network of residues mediating inter-domain contacts are shown in ball-and-stick representation. (C) Comparisons of the V-set IgSF domains of B7-1 and hCD2. (D) Two orthogonal views of domain 2 of B7-1 superimposed on the equivalent domains of hCD2. The C' or D strands are highlighted.

Immunoglobulin-like domains

The amino terminal domain (d1) of B7-1 has V-set topology, with β-strands forming DEB and AGFCC'C" β-sheets. Automated structure comparisons (DALI), rank B7-1d1 as most similar to the membrane distal domains of CD2 (1hnf) and CD4 (1cdy), molecules involved in cell-cell adhesion. Superimpositions of the B7-1 and CD2 V-set domains are shown in Fig. 1C (r.m.s. difference between B7-1d1 and CD2d1 is 1.1Å for 89 equivalent Cα atoms). An important structural similarity shared by B7-1d1 and CD2d1, distinguishing these domains from antigen receptor V-set domains, is the lack of overall twist in the AGFCC'C" sheet. This reflects the fact that B7-1d1 has its β-bulges located immediately after the CC' and FG turns at residues Met-38 and Arg-94, (equivalently positioned at residues Lys-43 and Asn-92 in human CD2d1; Bodian et al. 1994) whereas in antigen receptor V-set domains the β-bulges are located in the C' and G β-strands.

In contrast to d1, the membrane proximal domain of B7-1 (d2) exhibits greater similarity to the constant domains of antigen receptors and the membrane proximal domains of MHC antigen-type structures than to the membrane proximal domain of cell adhesion molecules. The d2 β-sandwich is formed by DEBA and GFC β-sheets typical of C1-set domains. Thus, the six structures most similar to B7-1d2 selected by DALI all contain C1-set domains, and structural superposition of B7-1d2 with the constant domain of immunoglobulin γ light chain reveals greater similarity than the superposition of B7-1d2 with CD2d2 (r.m.s. differences of 1.4 Å and 1.5 Å for 80 and 64 equivalent residues, respectively). The key structural feature is the organisation of the strands at the "edge" of the domain: whereas beyond the C strand CD2 has two short C'D strands extending the same β-sheet (Fig. 1D), the B7-1 and the Fab C1-set domains have extended DE strands forming the first half of the second β-sheet (Ikemizu et al. 2000).

Interdomain region

The interdomain "linkers" of B7-1 and human CD2 are identical in length (seven residues) and share remarkably similar main-chain conformations, even though only Pro-111 and Ile-113 are conserved in both sequences (B7-1 sequence numbering). After superimposition of the membrane proximal domains, the position of the membrane distal (NH2-terminal) domain of sB7-1 differs from that of sCD2 by a rotation of 100° about the long axis of the molecule and a reduction of 20° in tilt relative to the same axis. The rotation occurs about Ala-106 of B7-1 (Glu-104 in CD2). The reduced tilt, and the much longer D and E strands of d2, result in the interdomain region of B7-1 burying a larger surface area (670 vs 590 Å^2) than is the case for CD2. This region consists of a buried, hydrophobic core formed by Val-8, Pro-74, Ala-106, Phe-134 and Leu-163. In addition, electrostatic contacts and mainchain or sidechain hydrogen bonds between Ser-75 and Glu-162, and between Arg-73 and both Pro-159 and Glu-160, appear to stabilise the extended, upright stature of the extracellular region of B7-1. The elaborate network of interactions supporting d1 is shown in Fig. 1B.

The distinctive features of B7-1 define a new subset of the IgSF (Ikemizu et al. 2000). Although low, sequence identities indicate that these molecules have similar secondary and tertiary structures. The largest block of conserved residues that are not IgSF consensus residues lie in the D and E β-strands and DE loop of domain 2. The conservation of this region, which is involved in the key interdomain electrostatic contacts, along with the conserved hydrophobicity of the interdomain core (residues 8, 106, 134 and 163), suggests that the frameworks of these molecules are likely to be similar.

Ligand binding

Substantial losses (>90%) in ligand binding activity occur when the solvent exposed residues Arg-29, Tyr-31, Gln-33, Met-38, Ile-49, Trp-50 and Lys-86 of B7-1 are substituted with alanine (Fig. 2A; Peach et al. 1995). With the exception of Lys-86, these residues form a contiguous, L-shaped cluster identifying the d1 AGFCC'C" face as the ligand binding site of B7-1. Lys-86, which is not in the AGFCC'C" face, may nevertheless form a salt-bridge with Glu-24 in the BC loop that stabilises the conformation of the FG loop. Not all of the residues surrounding this patch have been mutated, however, and it is likely that the full binding surface is substantially larger than this. Gln-33 is completely conserved in human and mouse B7-1 and B7-2, as are aromatic residues at position 31 and hydrophobic residues at position 38. The conservation of these residues may explain the cross reactivities of human and mouse B7-1 and B7-2 with human and mouse CD28 and CTLA-4.

It has been proposed that d2 residues form part of the B7-1 ligand binding site (Peach et al. 1995). However, the relevant mutations (of DE loop residues Gln-157, Asp-158 and Glu-162) have much weaker effects on ligand binding (<50% inhibition) than mutations of the d1 AGFCC'C" face and inspection of the structure suggests that such effects could result from changes in the presentation of d1 via alterations in the network of interactions at the domain interface (Fig. 1B). Simple modelling also indicates that CTLA-4 cannot simultaneously bind both sets of residues. Furthermore, direct involvement of d2 would generate B7-1/ligand complexes spanning a much shorter inter-membrane distance (~10 nm) than TCR/MHC complexes (~15 nm), creating a steric barrier to the CTLA-4/TCR interactions required for inhibitory signaling by CTLA-4. For these reasons we conclude that the d1 AGFCC'C" face represents the principal ligand binding surface of B7-1.

The AGFCC'C" face of B7-1, is compositionally similar to other sites of protein-protein recognition and thus has substantially fewer charged and more hydrophobic residues than the equivalent face of CD2 which is also involved in low affinity cell surface recognition [29% charged residues versus 45-70% for

CD2 (Bodian et al. 1994)]. Surface electrostatic potential is depicted in Fig. 2B. Two hydrophobic residues essential for ligand binding, Met-38 and Trp-50, form a contiguous hydrophobic patch at the base of the AGFCC'C" sheet of d1. In a complementary manner, the conserved MYPPPY sequence motif implicated in the binding of both CTLA-4 and CD28 to B7-1 and B7-2 (Fig. 2D; Metzler et al. 1997), forms a hydrophobic patch centred on the FG loop of CTLA-4.

Manual docking of B7-1 d1 and CTLA-4, so that contact between the two hydrophobic patches and electrostatic complementarity are each maximised, and in a way that would permit interactions between adjacent cells, readily generated a feasible model for the complex of these two molecules (Ikemizu et al. 2000).

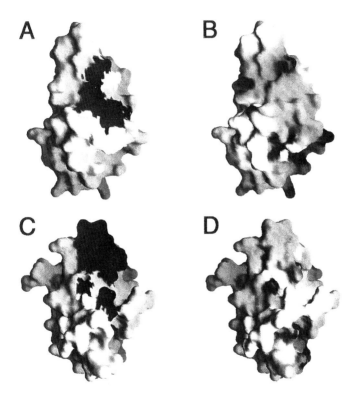

Fig. 2 Properties of the ligand binding faces of B7-1d1 (A, B) and CTLA-4 (C, D). In A and C, the GRASP surfaces of residues whose mutation to alanine disrupt or have no effect on binding (Peach et al. 1995; Metzler et al. 1997) are colored black and white, respectively. In B and D, the electrostatic potential is representated; black represents positive potential, gray represents neutral and white represents negative potential contoured at +8.5 kT. The line of view in each panel is essentially perpendicular to the AGFCC'C" ligand binding face of each molecule.

Dimerization of sB7-1

A striking and unexpected feature of the sB7-1 crystal lattice is that it is dominated by a molecular contact at the two fold axis which buries 610 Å^2 of surface area for each molecule. The lattice contact generates a compact dimer which, in overall shape and dimensions, is reminiscent of an antibody Fab' molecule (Fig. 3A). However, the dimer differs from an Fab' insofar as the d1 interaction is mediated by a relatively flat surface formed by the B, C'', D, and E β-strands rather than the AGFCC'C'' face and there is no contact between d2 of each molecule (although such contacts might be possible in vivo and may only be precluded by this particular lattice). The interaction is mediated by the primarily hydrophobic residues (Val-11, Val-22, Gly-45, Met-47, Ile-58, Asp-60, Ile-61, Thr-62 and Leu-70) decorating the B, C'', D and E β-strands, leaving the AGFCC'C'' face free for ligand binding. The molecules are inclined by 4° to the 2-fold axis, an arrangement consistent with the COOH-terminal stalks linking the molecules to the cell surface entering the membrane adjacently. The distribution of the glycosylation sites on the sB7-1 structure indicates that, even if all potential sites were to be fully utilised, the dimer interface would be free of potential N-linked glycans.

Dimerization of sB7-1 in solution was confirmed using analytical ultracentrifugation at a range of concentrations and at three temperatures. The results indicated that sB7-1 undergoes dimerisation in solution with a K_d in the range of 20-50 μM (Ikemizu et al. 2000); The temperature dependence of the K_d is consistent with the largely hydrophobic nature of the contact seen in the crystal lattice.

Cell-cell recognition

Functionally distinct interactions are suggested by the presence of radically different molecular contacts in the crystal lattices of sB7-1 and sCD2. sCD2 packs head-to-head, an arrangement which was predicted (Jones et al. 1992) to mimic natural receptor-ligand interactions. In contrast, the sB7-1 crystal lattice is dominated by a side-side molecular contact which generates a potentially bivalent homodimer. Model studies indicate that surface proteins with solution affinities in the range of 15-75 μM interact spontanously at the cell surface. The K_d of 20-50 μM for dimerisation and rapid monomer-dimer exchange we have observed suggests that B7-1 may exist at the cell surface in a dynamic equilibrium dominated by a homodimer similar to that observed in the crystal. Since CTLA-4 exists as a disulfide-bonded homodimer, the existence of B7-1 dimers would potentially allow two types of multivalent associations to form at the cell-cell interface, one involving paired homodimers (not shown), the alternative involving a B7-1 dimer bridging adjacent CTLA-4 dimers (Fig. 3B; Ikemizu et al. 2000). Simple modelling studies suggest that the second type of interaction might be sterically favoured. Recent work suggesting that mouse CTLA-4 forms a tight, rather than a loose, homodimer is also consistent with the arrangement shown in Fig. 3B (Ostrov et al. 2000).

CTLA-4 is not expressed on resting T cells and the level of CTLA-4 expression at the cell surface is kept very low, even on activated cells. CTLA-4 is also stored in intracellular vesicles whose release is directed at the contact zone between activated T cells and antigen presenting cells. This suggests that its release into, and sequestration within, the central supramolecular activation cluster (cSMAC) of the immunological synapse is key to the inhibitory function of CTLA-4. The solution affinity of B7-1 for CTLA-4 (~0.2-0.4 μM) is among the highest described for interacting cell surface molecules, at least 10 - fold higher than the affinity of B7-1/CD28 interactions (~4 μM; van der Merwe et al. 1997) and 40-100 - fold higher than that of B7-2/CD28 interactions (~15 μM). The avidity model proposed in Fig. 3B would provide a mechanism for further stabilising the interaction of B7-1 with CTLA-4 within the cSMAC. Whilst it is clear that B7-1/CD28 interactions may generate competing costimulatory signals, the foregoing structural considerations, along with the timing of B7-1 and CTLA-4 expression, suggest that a key function of B7-1 is to generate particularly stable signaling complexes with CTLA-4 late in immune responses, increasing the likelihood of inhibitory signaling by CTLA-4 and facilitating the termination of T-cell activation.

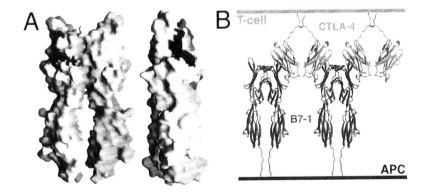

Fig. 3. Structure of the sB7-1 homodimer observed in the crystal lattice. (A) Two orthogonal views of the GRASP surface of the homodimer showing the location of residues whose mutation to alanine disrupts (black) or has no effect (white) on binding. (B) Potential molecular complexes formed by the B7-1 and CTLA-4 homodimers at the cell surface. The B7-1 homodimer shown is identical to that observed in the crystal lattice. The CTLA-4 homodimer is based on the solution structure of the monomer (Metzler et al. 1997), with the sequence connecting β-strand G to the intermolecular disulphide at Cys-123 modelled in extended conformation.

Acknowledgements

This work was supported by the Arthritis Research Campaign, the Biotechnology and Biological Sciences Research Council, the Engineering and Physical Sciences Research Council, the Medical Research Council, the Royal Society and The Wellcome Trust.

References

Bodian DL, Jones EY, Harlos K, Stuart DI, Davis SJ (1994) Crystal structure of the extracellular region of the human cell adhesion molecule CD2 at 2.5 Å resolution. Structure 15:755-766

Butters TD, Sparks LM, Harlos K, Ikemizu S, Stuart DI, Jones EY, Davis SJ (1999) Additive effects of N-butyldeoxynojirimycin and the Lec3.2.8.1 mutant phenotype on N-glycan processing in Chinese hamster ovary cells: application to glycoprotein crystallisation. Protein Sci 8:1696-1701

Ikemizu S, Gilbert RJC, Fennelly JA, Collins AV, Harlos K, Jones EY, Stuart DI (2000) Structure and dimerization of a soluble form of B7-1. Immunity 12:51-60

Jones EY, Davis SJ, Williams AF, Harlos K, Stuart DI (1992) Crystal structure at 2.8 Å resolution of a soluble form of the cell adhesion molecule CD2. Nature 360:232-239

Lenschow DJ, Walunas TL, Bluestone JA (1996) CD28/B7 system of T cell costimulation. Annu. Rev. Immunol. 14:233-258

May AP, Robinson RC, Aplin RT, Bradfield P, Crocker PR, Jones EY (1997) Expression, crystallization, and preliminary X-ray analysis of a sialicacid-binding fragment of sialoadhesin in the presence and absence of ligand. Protein Sci 6:717-721

McAdam AJ, Schweitzer AN, Sharpe AH (1998) The role of B7 co-stimulation in activation and differentiation of CD4+ and CD8+ T cells. Immunol Rev 165:231-247

Metzler WJ, Bajorath J, Fenderson W, Shaw SY, Constantine KL, Naemura J, Leytze G, Peach RJ, Lavoie TB, Mueller L, Linsley PS (1997) Solution structure of human CTLA-4 and delineation of a CD80/CD86 binding site conserved in CD28. Nat Struct Biol 4:527-531

Ostrov AD, Shi W, Schwartz JCD, Almo SC, Nathenson SG (2000) Structure of Murine CTLA-4 and its role in modulating T cell responsiveness. Science 290:816-819

Peach RJ, Bajorath J, Naemura J, Leytze G, Greene J, Aruffo A, Linsley PS (1995) Both extracellular immunoglobin-like domains of CD80 contain residues critical for binding T cell surface receptors CTLA-4 and CD28. J Biol Chem 8:21181-21187

van der Merwe PA, Bodian DL, Daenke S, Linsley P, Davis SJ (1997) CD80 (B7-1) binds both CD28 and CTLA-4 with a low affinity and very fast kinetics. J Exp Med 185:393-403

9
ITIM-bearing receptors in platelets

Daniel C. Snell, Jean-Max Pasquet and Steve P. Watson

Department of Pharmacology, University of Oxford, Mansfield Road, Oxford, OX1 3QT, UK

Summary. The immunoreceptor tyrosine-based activatory motif (ITAM) containing glycoprotein, GPVI, plays an important signalling role in platelet activation. GPVI binds to exposed collagen fibres at the site of vascular damage and the ensuing signalling events contribute to formation of a thrombus to repair the damaged area. The control of this activatory signalling pathway by immunoreceptor tyrosine-based inhibitory motif (ITIM) containing molecules is beginning to be understood in the platelet. Two such ITIM molecules are known in platelets. CD31 is expressed on human and mouse platelets, whilst PIR-B is expressed only on murine platelets. CD31 associates with the cytoplasmic phosphatases SHP-1 and SHP-2 and is phosphorylated upon platelet activation by collagen, thrombin and FcγRIIA cross-linking. Mice deficient in CD31 show enhanced responses to GPVI agonists, demonstrating that CD31 is a negative regulator of platelet-collagen responses. PIR-B has recently been shown to be expressed on murine platelets and megakaryocytes. PIR-B is constitutively phosphorylated in murine platelets and associates with unidentified tyrosine phosphorylated proteins of 80, 73 and 38 kDa upon activation by thrombin or GPVI. In megakaryocytes, there is minimal phosphorylation of PIR-B under basal or activated conditons. No functional inhibitory response for PIR-B has been identified in platelets and no phosphatases have been associated with this molecule in platelets. The role for inhibitory receptor systems in platelets to control collagen-induced activation is still in its infancy and further work in this field is rapidly progressing.

Key words. Platelets, CD31, PIR-B, GPVI, Collagen

Introduction

Platelets are major blood components which adhere to exposed subendothelial matrix proteins, including collagen fibres, at the site of vascular damage and thus form a monolayer of cells over the damaged area. Additionally, upon binding to collagen fibres, platelets become activated, promoting platelet-platelet interaction and coagulant activity.

There are a number of proposed collagen receptors on the platelet surface, of which the integrin GPIa-IIa ($\alpha2\beta1$) and glycoprotein VI (GPVI) are generally accepted as the most critical in the processes of adhesion and activation, respectively (for review see Watson and Gibbins 1998). GPVI is a 60-65 kDa type I transmembrane glycoprotein and is a member of the immunoglobulin superfamily (Clemetson et al. 1999). GPVI is known to associate at the cell surface with the Fc receptor γ-chain in human and murine platelets (Nieswandt et al. 2000; Gibbins et al. 1997; Tsuji et al. 1997). Upon binding of collagen to GPVI, tyrosine phosphorylation of the Immunoreceptor Tyrosine-based Activation Motif (ITAM) of FcR γ-chain takes place by the Src family kinases, Fyn and Lyn (Ezumi et al. 1998; Briddon and Watson 1999; Melford et al. 1997). The ensuing signalling pathway involves a cascade of phosphorylation of adaptor and effector molecules via a system similar to that used by immunoreceptors and with a critical role for the tyrosine kinase Syk (Poole et al. 1997).

The regulation of immune responses in many cell types has been shown to be reliant on the role of inhibitory receptor systems which maintain the equilibrium between activatory and inhibitory signals (for review see Ravetch and Lanier 2000). Inhibitory receptors signal via immunoreceptor tyrosine-based inhibition motifs (ITIMs). It is therefore important to consider whether such inhibitory receptors are expressed on platelets, and whether they play a role in regulating GPVI dependent platelet activation.

CD31

The ITIM containing receptor, Platelet Endothelial Cell Adhesion Molecule-1 (PECAM-1/CD31) has been identified on platelets. CD31 is a 130 kDa glycoprotein which is expressed on the surface of platelets, endothelial cells, monocytes, neutrophils and subsets of T-lymphocytes (Ohto et al. 1985; Muller et al. 1989; Albelda et al. 1990; Newman et al. 1990). The intracellular domain of CD31 contains two ITIMs which, when phosphorylated, associate with the cytoplasmic phosphatases SHP-1 and SHP-2 (Jackson et al. 1997; Hua et al. 1998). The group of Newman has recently shown that attenuation of Ca^{++} mobilisation by CD31 is dependent on SHP-2, whilst SHP-1 is less important (Newman et al. 2000). It has been shown that activation of platelets through GPVI and by thrombin receptors results in tyrosine phosphorylation of CD31 (Jackson et al. 1997; Cicmil et al. 2000a). FcγRIIA is a low affinity Fc receptor on platelets, and it has also been shown that crosslinking of this receptor results in CD31 tyrosine phosphorylation (Noda et al. 2000). FcγRIIA and the GPVI/Fc γ-chain receptor complex are the only known ITAM containing receptors on human platelets. CD31 is thought to be phosphorylated by the Src family kinases, since it co-immunoprecipitates with Fyn, Lyn, Src, Yes and Hck in human platelets (Cicmil et al. 2000a). Recently, the functional role of CD31 in platelets has been investigated. The group of Gibbins has shown that cross-linking of CD31 on the platelet surface results in reduced platelet aggregation to collagen and to the GPVI selective agonist convulxin (Cicmil et al. 2000b). Similarly, through the use of CD31 deficient mice, Newman's group reports exaggerated platelet responses in these mice compared to wild type littermates. These responses include increased adhesion to immobilised fibrillar collagen, increased collagen-induced platelet aggregation and increased GPVI/Fcγ-chain induced dense granule secretion (Patil et al. 2000). Therefore it is clear that CD31 acts as a negative regulator of platelet-collagen interactions.

PIR-B

We have recently identified the presence of the ITIM-containing glycoprotein Paired Immunoglobulin-like Receptor-B (PIR-B) on murine platelets and megakaryocytes. PIRs are type I transmembrane glycoproteins which are expressed on haematopoietic cells, including B lymphocytes and mast cells. PIR-B contains six conserved Ig-like domains followed by two alternative cytoplasmic sequences giving A and B isoforms (Kubagawa et al. 1997). PIR-A has a short cytoplasmic tail and a charged arginine residue in its transmembrane domain enabling it to associate with the FcR γ-chain (Maeda et al. 1998; Kubagawa et al. 1999; Ono et al. 1999). In contrast, PIR-B has a long cytoplasmic tail that contains four ITIMs. PIR-B has been shown to selectively inhibit signalling by immune receptors through recruitment of SHP-1 and SHP-2 (Kubagawa et al. 1999; Hayami et al., 1997; Blery et al., 1998; Maeda et al., 1998). In B lymphocytes, the inhibitory effect of SHP-1 is mediated primarily by dephosphorylation of the tyrosine kinases Syk and Btk (Maeda et al. 1999). The ligands for PIRs are at this time unknown, however it is speculated that PIRs bind MHC class I-related proteins. This is analogous to the related killer cell inhibitory receptors (KIRs).

We have shown that PIR-B is constitutively tyrosine phosphorylated in murine platelets and associates with unidentified tyrosine phosphorylated proteins of 80, 73 and 38 kDa upon activation by thrombin or the collagen receptor GPVI. In murine megakaryocytes, there is minimal phosphorylation of PIR-B under basal or activated conditions. Crosslinking of PIR-B on the surface of platelets did not alter the phosphorylation state of PIR-B nor did it inhibit GPVI-dependent platelet responses. We could not identify an association between PIR-B and SHP-1 or SHP-2 in murine platelets or megakaryocytes. Therefore the inhibitory role of

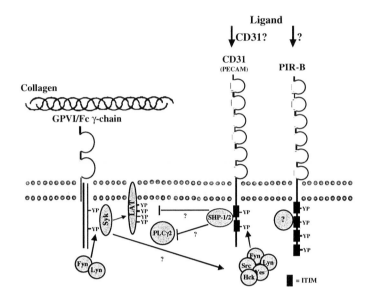

Fig. 1. Schematic characterisation of CD31 and PIR-B roles in GPVI signalling. GPVI binds to collagen and tyrosine phosphorylation of the Fc γ-chain takes place. CD31 is phosphorylated upon GPVI activation, possibly by Src family kinases. This leads to the recruitment of SHP-1 and SHP-2 and the subsequent inhibition of GPVI-specific responses by unknown mechanisms. The role of PIR-B is less well characterised. YP = phosphorylated tyrosine residue.

PIR-B is at this time unclear, however it is possible that another, as yet unidentified, phosphatase may have a role in the function of PIR-B.

Conclusions

The role of the ITIM containing molecule CD31 in controlling activatory pathways in platelets is beginning to emerge. It seems likely that other ITIM containing molecules exist on platelets and that each may have distinct roles in platelet function.

Acknowledgements

This work was supported by the British Heart Foundation. SPW is a BHF senior Research Fellow.

References

Albelda SM, Oliver PD, Romer LM, Buck CA (1990) EndoCAM: a novel endothelial cell-cell adhesion molecule. J Cell Biol 110:1227-1237

Blery M, Kubagawa H, Chen C-C, Vely F, Cooper MD, Vivier E (1998) The paired Ig-like receptor PIR-B is an inhibitory receptor that recruits the protein-tyrosine phosphatase SHP-1. Proc Natl Acad Sci U S A. 95: 2446-2451.

Briddon SJ, Watson SP (1999) Evidence for the involvement of p59fyn and p53/56lyn in collagen receptor signalling in human platelets. Biochem J 338: 203-209

Cicmil M, Thomas JM, Sage T, Barry FA, Leduc M, Bon C, Gibbins JM (2000a) Collagen, convulxin, and thrombin stimulate aggregation-independent tyrosine phosphorylation of CD31 in platelets. Evidence for the involvement of Src family kinases. J Biol Chem 275: 27339-27347

Cicmil M, Thomas JM, Sage T, Barry FA, Gibbins JM (2000b) PECAM-1: A negative regulator of platelet activation? Blood 96:245a

Clemetson JM, Polgar J, Magnenat E, Wells TN, Clemetson KJ (1999) The platelet collagen receptor glycoprotein VI is a member of the immunoglobulin superfamily closely related to FcalphaR and the natural killer receptors. J Biol Chem 274: 29019-29024

Ezumi Y, Shindoh K, Tsuji M, Takayama H (1998) Physical and functional association of the Src family kinases Fyn and Lyn with the collagen receptor glycoprotein VI-Fc receptor gamma chain complex on human platelets. J Exp Med 188: 267-276

Gibbins JM, Okuma M, Farndale R, Barnes M, Watson SP (1997) Glycoprotein VI is the collagen receptor in platelets which underlies tyrosine phosphorylation of the Fc receptor gamma-chain. FEBS Lett 413: 255-259

Hayami K, Fukuta D, Nishikawa Y, Yamashita Y, Inui M, Ohyama Y, Hikida M, Ohmori H, Takai T (1997) Molecular cloning of a novel murine cell-surface glycoprotein homologous to killer cell inhibitory receptors. J Biol Chem 272: 7320-7327

Hua CT, Gamble JR, Vadas MA, Jackson DE (1998) Recruitment and activation of SHP-1 protein-tyrosine phosphatase by human platelet endothelial cell adhesion molecule-1 (PECAM-1). Identification of immunoreceptor tyrosine-based inhibitory motif-like binding motifs and substrates. J Biol Chem 273: 28332-28340

Jackson DE, Ward CM, Wang R, Newman PJ (1997) The protein-tyrosine phosphatase SHP-2 binds platelet/endothelial cell adhesion molecule-1 (PECAM-1) and forms a distinct signaling complex during platelet aggregation. Evidence for a mechanistic link between PECAM-1- and integrin-mediated cellular signaling. J Biol Chem. 272: 6986-6993

Kubagawa H, Burrows PD, Cooper MD (1997) A novel pair of immunoglobulin-like receptors expressed by B cells and myeloid cells. Proc.Natl.Acad.Sci.USA 94: 5261-5266

Kubagawa H, Chen CC, Ho LH, Shimada TS, Gartland L, Mashburn C, Uehara T, Ravetch JV, Cooper MD (1999) Biochemical nature and cellular distribution of the paired immunoglobulin-like receptors, PIR-A and PIR-B. J Exp Med 189: 309-318

Maeda A, Kurosaki M, Ono M, Takai T, Kurosaki T (1998) Requirement of SH2-containing protein tyrosine phosphatases SHP-1 and SHP-2 for paired immunoglobulin-like receptor B (PIR-B)-mediated inhibitory signal. J Exp Med 187: 1355-1360

Maeda A, Scharenberg AM, Tsukada S, Bolen JB, Kinet JP, Kurosaki T (1999) Paired immunoglobulin-like receptor B (PIR-B) inhibits BCR-induced activation of Syk and Btk by SHP-1. Oncogene 18: 2291-2297

Melford SK, Turner M, Briddon SJ, Tybulewicz VL, Watson SP (1997) Syk and Fyn are required by mouse megakaryocytes for the rise in intracellular calcium induced by a collagen-related peptide. J Biol Chem 272: 27539-27542

Muller WA, Ratti CM, McDonnell SL, Cohn ZA (1989) A human endothelial cell-restricted, externally disposed plasmalemmal protein enriched in intercellular junctions. J Exp Med 170: 399-414

Newman PJ, Berndt MC, Gorski J, White GC, Lyman S, Paddock C, Muller WA (1990) PECAM-1 (CD31) cloning and relation to adhesion molecules of the immunoglobulin gene superfamily. Science 247: 1219-1222

Newman DK, Hamilton CA, Armstrong MJ, Newman PJ (2000) Inhibition of antigen-receptor signaling by platelet endothelial cell adhesion molecule-1 (CD31) requires an intact ITIM, SHP-2, and p56^{lck}. Blood 96:238a

Nieswandt B, Bergmeier W, Schulte V, Rackebrandt K, Gessner JE, Zirngibl H (2000) Expression and function of the mouse collagen receptor glycoprotein VI is strictly dependent on its association with the FcRgamma chain. J Biol Chem 275: 23998-24002

Noda M, Paddock CM, Liu C-Y, Huang Z, Schreiber AD, Newman DK, Newman PJ (2000) Regulation of FcγRIIA-mediated phagocytosis by the ITIM-bearing inhibitory receptor, PECAM-1. Blood 96:243a

Ohto H, Maeda H, Shibata Y, Chen RF, Ozaki Y, Higashihara M, Takeuchi A, Tohyama H (1985) A novel leukocyte differentiation antigen: two monoclonal antibodies TM2 and TM3 define a 120-kd molecule present on neutrophils, monocytes, platelets, and activated lymphoblasts. Blood 66: 873-881

Ono M, Yuasa T, Ra C, Takai T (1999) Stimulatory function of paired immunoglobulin-like receptor-A in mast cell line by associating with subunits common to Fc receptors. J Biol Chem 274: 30288-30296

Patil S, Newman DK, Newman PJ (2000) PECAM-1 serves as an inhibitory receptor that modulates platelet responses to collagen. Blood 96:446a

Poole A, Gibbins JM, Turner M, van Vugt MJ, van de Winkel JG, Saito T, Tybulewicz VL, Watson SP (1997) The Fc receptor γ-chain and the tyrosine kinase Syk are essential for activation of mouse platelets by collagen. EMBO J 16: 2333-2341

Ravetch JV, Lanier LL (2000) Immune inhibitory receptors. Science 290: 84-89

Tsuji M, Ezumi Y, Arai M, Takayama H (1997) A novel association of Fc receptor gamma-chain with glycoprotein VI and their co-expression as a collagen receptor in human platelets. J Biol Chem 272: 23528-23531

Watson SP, Gibbins J (1998) Collagen receptor signalling in platelets: Extending the role of the ITAM. Immunol Today 19: 260-264

10
Identification of Fcα/μ receptor expressed on B lymphocytes and macrophages

Akira Shibuya[1,2], Kazuko Shibuya[1], Yoshio Shimizu[1], Katsumi Yotsumoto[1], and Hiromitsu Nakauchi[1,3]

[1]Department of Immunology, Institute of Basic Medical Sciences, University of Tsukuba, [2]TOREST (JST) and [3]CREST (JST), 1-1-1 Tennodai, Tsukuba, Ibaraki 305-8575, Japan

Summary. IgM is the first antibody to be produced in a humoral immune response and plays an important role in the primary stage of immunity. We have identified a novel Fc receptor for IgM and IgA, designated Fcα/μ receptor (Fcα/μR), which is a new member of Fc receptor on the chromosome 1 gene family. The Fcα/μR is constitutively expressed on the majority of B-lymphocytes and macrophages. The Fcα/μR mediates endocytosis *Staphylococcus aureus* /anti-S. aureus IgM antibody immune complexes by primary B-lymphocytes. These results reveal a new mechanism in the primary stage of immune defense against microbes.

Key Words. IgM, IgA, Fc receptor, Endocytosis, B lymphocytes

Introduction

IgM is the first antibody to be produced in a humoral immune response. In addition, natural antibodies, involved in the prevention of pathogen dissemination to vital organs, are mainly IgM (Ochsenbein et al. 1999). Therefore, IgM may play an important role in the innate stage of immunity. However, it is not known how IgM protects against infection.

The antibodies or immune complexes bind to cell surface Fc receptors on hematopoietic cells and play important roles in a wide array of immune responses, including antibody-dependent cellular cytotoxicity, mast cell degranulation, phagocytosis, cell proliferation, antibody secretion and enhancement of antigen presentation (Ravetch and Clynes 1998). Functional Fc receptors for IgM have been reported on subpopulations of human and rodent T, B and NK cells (Whelan 1981; Ercolani et al. 1981; Mathur et al. 1988; Ohno et al. 1990; Nakamura et al. 1993; Pricop et al. 1993). In contrast to the well-defined Fcγ and Fcε receptors, a gene encoding a Fcμ receptor has not previously been identified, although a polymeric Ig receptor able to bind IgM and IgA has been described (Mostov 1994). Similarly, although a human Fcα receptor (*CD89*) gene has been reported (Maliszewski et al 1990), a homologous rodent Fcα receptor has not yet been found. We have recently cloned a novel Fc receptor for IgM and IgA, designated Fcα/μR (Shibuya et al. 2000). Here, we describe a molecular and functional characteristics of Fcα/μR.

Molecular cloning and characterization of the Fcα/μR

To identify a Fc receptor for IgM, we screened COS-7 cells transfected with a cDNA library prepared from the mouse T cell leukemia BW5147 by using a mouse IgM monoclonal antibody (mAb) as a probe. We isolated a cDNA clone of 2,361 bp that contains an open reading frame encoding a type I transmembrane protein (Genbank Accession Number; AB048834) (Shibuya et al. 2000). A pair of cysteine residues in the extracellular domain was flanked by the consensus sequence for Ig-like domains, indicating that this molecule, designated mFcα/μR, is a member of the Ig supergene family. A human cDNA with homology to the mFcα/μR (hFcα/μR) was identified in a database (Genbank Accession Number; E15470) (Shibuya et al. 2000) and was found to encode a protein with 49% amino-acid identity to the mFcα/μR. These genes were mapped by FISH to syntenic regions of mouse chromosome 1 (1F) and human chromosome 1 (1q32.3), near several other Fc receptors, including Fcγ receptors I, II, and III, Fcα receptor, and the polymeric Ig receptor (Fig. 1). The Fcα/μR is unique, demonstrating no significant homology to other proteins or nucleotide sequences in databases. However, we observed a motif in the Ig-like domain conserved in the first Ig-like domain of the human, bovine, and rodent polymeric Ig receptor, which binds to the Fc of IgA and IgM (Fig. 2). Overall, the Fcα/μR has less than 10% amino acid homology with the mouse and human polymeric Ig receptor. Immunoprecipitation of the Flag-tagged mFcα/μR expressed on the COS-7 transfectants with an anti-Flag mAb revealed an ~70 kDa protein, when analyzed using both reducing and non-reducing conditions (Shibuya et al. 2000).

Expression of the Fcα/μR

RT-PCR analysis demonstrated that the mFcα/μR transcripts is expressed in several tissues, including thymus, spleen, liver, kidney, small and large intestines, testis, and placenta (Shibuya et al. 2000). Northern blot analysis further demonstrated that the mFcα/μR transcripts is abundantly expressed in kidney, small intestine and lymph node (Sakamoto et al., submitted).

To confirm the cell surface expression of the mFcα/μR protein, we generated a monoclonal antibody (rat IgG isotype) against the mFcα/μR. Analysis of spleen cells by flow cytometry showed that the mFcα/μR is expressed on the majority of B cells and macrophages, but not on granulocytes, T cells or NK cells (Shibuya et al. 2000). We compared the mFcα/μR expression between IgD⁻, IgM⁺ immature and IgD⁺, IgM⁺ mature B lymphocytes. While IgD⁺ mature B lymphocytes expressed significant amount of the mFcα/μR, it was not expressed on the majority of IgD⁻ immature B lymphocytes (Sakamoto, et al., submitted). We also found that pre-B cell tumor line Ba/F3 did not express the mFcα/μR. These results suggest that the expression of the mFcα/μR is developmentally regulated and that the mFcα/μR may have functional role in some aspect of B cell maturation and activation.

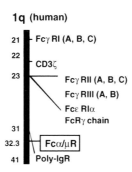

Fig. 1. The human Fcα/μR gene was mapped by FISH to chromosome 1 (1q32.3), near several other Fc receptors, including Fcγ receptors I, II, and III, Fcα receptor, and the polymeric Ig receptor. The mouse Fcα/μR gene was mapped to syntenic region of mouse chromosome 1 (1F).

Fig. 2. Comparison of the amino acid sequences of the predicted binding sites for IgA and IgM of the Fcα/μR and the polymeric Ig receptors. The predicted binding sites in the Ig-like domain of the Fcα/μR is conserved in the first Ig-like domain of the human, bovine, and rodent polymeric Ig receptor, which binds to the Fc of IgA and IgM. However, overall, the Fcα/μR has less than 10% amino acid homology with the mouse and human polymeric Ig receptor.

Binding of IgM and IgA to the mFcα/μR

The Ba/F3 transfectant stably expressing the mFcα/μR specifically bound several mouse IgM mAbs and both dimeric and monomeric IgA mAbs, but not any IgG mAbs tested (including mouse IgG1, IgG2a, IgG2b and IgG3). By contrast, the parental Ba/F3 cells, which do not express the mFcα/μR, did not bind any of these mAbs. Moreover, the mFcα/μR transfectant bound the Fc, but not F(ab')$_2$, fragment of human IgM (Shibuya et al. 2000). Scatchard analysis using ^{125}I-labeled IgM and IgA demonstrated that mFcα/μR expressed on Ba/F3 transfectant bound IgM and IgA with high or intermediate affinity (Shibuya et al. 2000). However, while primary mature B-lymphocytes express sufficient amount of the mFcα/μR, binding of IgM and IgA to B lymphocytes were barely detectable, as determined by flow cytometry (Sakamoto, et al., submitted), suggesting that the affinity of the mFcα/μR expressed on resting B lymphocytes for IgM and IgA is low. However, binding of both IgM and IgA to primary B lymphocytes was significantly increased after culture in medium supplemented with 10% FBS (Sakamoto, et al., submitted). These results suggest that the affinity of the mFcα/μR on primary B lymphocytes for IgM and IgA was increased after culture, in which cellular interaction between B lymphocytes and the other spleen cells, cytokines produced from these cells, and /or exogenous factors in FBS might have been involved.

Endocytosis of IgM-coated micro particles mediated by the mFcα/μR

The cytoplasmic domain of the mFcα/μR contains a di-leucine motif at residues 519 and 520, which has been implicated in endosome and lysosome targeting of diverse proteins and is involved in agonist-induced internalization (Aiken et al. 1994; Hamer et al. 1997; Gabilondo et al. 1997; Craig et al. 1998) A di-leucine motif is involved in FcγRIIB-mediated endocytosis (Hunziker and Fumey 1994). To examine whether the mFcα/μR is involved in endocytosis, Ba/F3 transfectants expressing the mFcα/μR were incubated with IgM-coated fluorescent beads at 37°C. After extensive washing and treatment with trypsin, we analyzed the cells by flow cytometry, which revealed a subpopulation of the Ba/F3 transfectants containing fluorescent beads by flow cytometyry (Shibuya et al. 2000). The fluorescent beads were also detected in the cytoplasm of the transfectant using immunofluorescent microscopy and confocal scanning laser microscopy, indicating that the mFcα/μR was involved in endocytosis of IgM-coated beads. However, the fluorescent beads were not detected in any of the Ba/F3 transfectants expressing mFcα/μR mutated (L-A) at residue 519, 520 or both (Shibuya et al. 2000). These results indicate that both leucines at residue 519 and 520 are required for endocytosis by the mFcα/μR. The human mFcα/μR does not have a di-leucine motif in its cytoplasmic domain; we are investigating whether other elements in the human receptor might also serve this function.

To examine the biological significance of the mFcα/μR expressed on primary B-lymphocytes, mouse spleen cells were incubated with immune complexes composed of FITC-labeled *S. aureus* bacteria coated with IgM or IgG anti-*S. aureus* mAbs. B220⁺ spleen cells cultured with the immune complex of IgM anti-*S. aureus* and FITC-labeled *S. aureus* bacteria contained one or more FITC-labeled *S. aureus* bacteria in the cytoplasm, as determined by immunofluorescent microscopy. By contrast, IgG immune complex of *S. aureus* was not detected in B220⁺ spleen cells (Shibuya et al. 2000). These results indicate that the mFcα/μR mediates endocytosis of IgM-coated microbial pathogens.

Concluding remarks

The Fcα/μR is a new member of the chromosome 1 gene family of Fc receptors (Fig. 1). Because the Fc receptors on the chromosome show the rich diversity of structures and functions, one raises a question whether the Fcα/μR is a single-family member of Fc receptor that binds IgM and IgA (Kinet and Launay 2000). Although southern blot analysis of mouse genomic DNA suggested that the Fcα/μR is a single gene (Shimizu, et al., unpublished observation), the presence of an alternate splicing form or the products of supplementary genes cannot be excluded. In addition, previous reports described other Fc receptors for IgM, including a glycosyl phosphatidylinosytol linked protein expressed on B cells (Ohno et al. 1990) and a protein on T cells (Nakamura et al. 1993). Future studies on the chromosome should resolve this issue.

The Fcα/μR is expressed on the majority of B lymphocytes and macrophages. Moreover, mFcα/μR is also expressed in kidney, intestine and lymph node. At present, it is unclear which cell type in these hematopoietic and non-hematopoietic tissues specifically expresses the Fcα/μR. Nonetheless, because the Fcα/μR mediates the endocytosis of IgM-coated bacteria by B-lymphocytes, the Fcα/μR may play an important role for immunity in these organs. Recent reports demonstrated that mice deficient in the secretory form of IgM exhibit delayed development of specific IgG antibodies to T cell-dependent foreign antigens (Ehrenstein et al. 1998) and dissemination of micropathogens in peripheral organs but not in secondary lymphoid organs (Ochsenbein et al. 1999), suggesting that the Fcα/μR may play an essential role in the priming of helper T lymphocytes in secondary lymphoid organs and in immune defense against bacterias in peripheral organs.

References

Aiken C, Konner J, Landau NR, Lenburg ME, Trono D (1994) Nef induces CD4 endocytosis: requirement for a critical dileucine motif in the membrane-proximal CD4 cytoplasmic domain. Cell 76:853-864

Craig HM, Pandori MW, Guatelli JC (1998) Interaction of HIV-1 Nef with the cellular dileucine-based sorting pathway is required for CD4 down-regulation and optimal viral infectivity. Proc Natl Acad Sci USA 95:11229-11234

Ehrenstein MR, O'Keefe TL, Davies SL, Neuberger MS (1998) Targeted gene disruption reveals a role for natural secretory IgM in the maturation of the primary immune response. Proc Natl Acad Sci USA. 95: 10089-10093

Ercolani L, Novogrodsky A, Suthanthiran M, Rubin AL, Stenzel KH (1981) Expression of Fc mu receptors by human T lymphocytes: effects of enzymes, metabolic inhibitors, and X-irradiation. J Immunol 127:2044-2051

Gabilondo AM, Hegler J, Krasel C, Boivin-Jahns V, Hein L, Lohse MJ (1997) A dileucine motif in the C terminus of the β2-adrenergic receptor is involved in receptor internalization. Proc Natl Acad Sci USA 94:12285-12290

Hamer I, Haft CR, Paccaud JP, Maeder C, Taylor S, Carpentier JL (1997) Dual role of a dileucine motif in insulin receptor endocytosis. J Biol Chem 272:21685-21691

Hunziker W. and Fumey C. (1994) A di-leucine motif mediates endocytosis and basolateral sorting of macrophage IgG Fc receptors in MDCK cells. EMBO J 13:2963-2967

Kinet J-P, Launay, P (2000) Fcα/μR: single member or first born in the family? Nature Immunol 1:371-372

Maliszewski CR, March CJ, Schoenborn MA, Gimpel S, Shen L. (1990) Expression cloning of a human Fc receptor for IgA. J Exp Med 172:1665-1672

Mathur A, Lynch RG, Kohler G. (1998) Expression, distribution and specificity of Fc receptors for IgM on murine B cells. J Immunol 141:1855-1862

Mostov KE (1994) Transepithelial transport of immunoglobulins. Annu Rev Immunol 12:63-84

Nakamura T, Kubagawa H, Ohno T, Cooper MD (1993) Characterization of an IgM Fc-binding receptor on human T cells. J Immunol 151:6933-6941

Ochsenbein AF, Fehr T, Lutz C, Suter M, Brombacher F, Hengartner H, Zinkernagel RM (1999) Control of early viral and bacterial distribution and disease by natural antibodies. Science 286:2156-2159

Ohno T, Kubagawa H, Sanders SK, Cooper MD (1990) Biochemical nature of a Fcμ receptor on human B-lineage cells. J Exp Med 172:1165-1175

Pricop L, Rabinowich H, Morel PA, Sulica A, Whiteside TL, Herberman RB (1993) Characterization of the Fcμ receptor on human natural killer cells. Interaction with its physiologic ligand, human normal IgM, specificity of binding, and functional effects. J Immunol 151:3018-3029

Ravetch JV, Clynes RA (1998) Divergent roles for Fc receptors and complement *in vivo*. Annu Rev Immunol 16:421-432

Shibuya A, Sakamoto N, Shimizu Y, Shibuya K, Osawa M, Hiroyama T, Eyre HJ, Sutherland GR, Endo Y, Fujita T, Miyabayashi T, Sakano S, Tsuji T, Nakayama E, Phillips JH, Lanier LL, Nakauchi H (2000) Fcα/μ receptor mediates endocytosis of IgM-coated microbes. Nature immunol 1:441-446

Whelan CA (1981) A functional role for Fc μ receptors on human lymphocytes. Immunol Lett 3:249-254

11

Functional characterization of mouse CD94 by using a novel monoclonal antibody

Noriko Toyama-Sorimachi[1,2], Hideo Yagita[3], Fujiko Kitamura[2], Akemi Kawasaki[3], Shigeo Koyasu[4], and Hajime Karasuyama[1,2]

[1]Department of Immune Regulation, Tokyo Medical and Dental University, Graduate School, 1-5-45 Yushima, Bunkyo-ku, Tokyo 113-8510, Japan
[2]Department of Immunology, Tokyo Metropolitan Organization for Medical Research, The Tokyo Metropolitan Institute of Medical Science, 3-18-22 Hon-komagome, Bunkyo-ku, Tokyo 113-8613, Japan
[3]Department of Immunology, Juntendo University School of Medicine, 2-1-1 Hongo, Bunkyo-ku, Tokyo 113-8421, Japan
[4]Department of Microbiology and Immunology, Keio University School of Medicine, 35 Shinanomachi, Shinjuku-ku, Tokyo 160-8582, Japan

Summary. Cytotoxic activity of NK cells is regulated in delicate balance between activating signals and inhibitory signals mediated by functionally different receptor families, so-called activating receptors and inhibitory receptors. With the aim of identifying novel NK receptors, we generated mAbs by immunizing activated NK cells and selected a novel mAb designated Yuri3. Biochemical and molecular analysis indicated that Yuri3 recognizes mouse CD94. Here we explored the expression of mouse CD94 and evaluated its function on primary NK cells by using the mAb. CD94 was expressed on essentially all NK and NKT cells as well as some subsets of T cells. Two distinct populations were observed in NK and NKT cells, CD94^{bright} and CD94^{dull} cells, and the expression of CD94 was independent of the expression of Ly-49 family members. Addition of the anti-CD94 mAb abrogated the inhibition of the target cell lysis mediated by the NK recognition of Qa-1. Interestingly, CD94^{bright} but not CD94^{dull} population was responsible for Qa-1 recognition. Substantial cytotoxicity against autologous target cells as well as enhanced cytotoxicity against allogeneic target cells was observed in the presence of the mAb, suggesting that CD94 plays an important role in self protection from NK cytotoxicity and its inhibitory function is independent of inhibitory receptors for classical MHC class I such as Ly-49 family members.

Key words. NK cells, CD94, Antibody, Qa-1, Inhibitory receptor

Introduction

Inhibitory receptors on NK cells recognize MHC class I molecules on target cells, and signals through the inhibitory receptors turn off NK-cell mediated cytotoxicity (Lanier 1998). Recently a number of receptor molecules on NK cells have been identified in both human and rodent (Colonna 1997; Lanier 1998; Long and Wagtmann 1997; Moretta et al. 1997; Takei et al. 1997; Yokoyama 1993). In order to identify novel NK receptors, we have been generating mAbs by immunizing NK cells and recently established a novel mAb designated Yuri3. Biochemical and molecular analysis indicated that Yuri3 recognizes mouse CD94. CD94 is one of members of lectin-like receptor family and paired with some members of NKG2 family (Lopez-Botet et al. 1998). Most important function of CD94/NKG2 on human NK cells is recognition of HLA-E on target cells (Braud and McMichael 1999). HLA-E is one of non-classical MHC class I molecules and appears to associate predominantly with peptides derived from leader sequences of MHC class I in TAP-dependent manner. As a result, NK cells can monitor the biosynthesis of MHC class I as well as TAP function on target cells. Here we employed a novel mAb specific for mouse CD94 to explore the expression of mouse CD94 and evaluate its function on primary NK cells.

Tissue distribution of mouse CD94

Mouse CD94 was expressed on essentially all NK and NKT cells in both spleen and thymus. Two distinct populations were identified with regard to the expression level of CD94, CD94^{bright} and CD94^{dull}. In C57BL/6 spleen, CD94^{bright} and CD94^{dull} populations were detectable at the similar frequency (about 50% each). In the thymus, approximately a quarter of NKT cells were CD94^{bright} and the rest of cells were CD94^{dull}. CD94 was expressed on DX5$^+$ NK cells not only from C57BL/6 mice but also from various inbred mouse strains including BALB/c, C3H, DBA2 and AKR. Both CD94^{bright} and CD94^{dull} populations were observed in all strains tested, though levels of CD94 expression and proportion of CD94^{bright} and CD94^{dull} populations varied among different strains.

Besides NK and NKT cells, a small proportion of splenic T cells was found to express CD94, and those T cells were CD4$^-$CD8$^+$ and CD4$^-$CD8$^-$. CD94$^+$ cells were also detected in both αβT and γδT cell subsets among intestinal IELs. In contrast, Ly-49A$^+$ cells and Ly-49C/I$^+$ cells were observed predominantly in αβT cells. Thus, the usage of NK receptors appears to be different in these two subsets of T cells.

Expression of CD94 was detected among NK1.1$^+$ cells in the fetal liver as early as on day 14 of gestation, that is, before the expression of known Ly-49 family members (Raulet et al. 1997). This is consistent with previous observation that Qa-1 binding cells exist in fetal NK cells (Sivakumar et al. 1999; Toomy et al. 1999). Proportion of CD94[+] cells among fetal liver NK cells increased during ontogeny. In fetal liver NK cells, there was no clear discrimination of CD94[bright] and CD94[dull] populations unlike in adult NK cells. Taken together with our results and previous observation, it is suggested that at least some of CD94 molecules expressed on fetal NK cells are functional in recognition of Qa-1. Therefore, CD94 may play some roles in the immunological regulation during development.

Effect of MHC class I on CD94 expression

Previous studies demonstrated that the expression of NK receptors such as Ly-49 family members is influenced by the expression of their ligands, MHC class I molecules (Olsson et al. 1995; Salcedo et al. 1996). We found that proportion of the CD94[bright] population among NK cells was significantly higher in C57BL/6 β2m-/-mice than in C57BL/6 mice. Levels of CD94 expression in both CD94[bright] and CD94[dull] populations were slightly higher in C57BL/6 β2m-/- mice. These results suggest that the expression of CD94 on NK cells could be regulated by the expression of β2m-associated molecules.

Involvement of CD94 in Qa-1 recognition by NK cells

In our cytotoxic assay using CHO target cells stably transfected with Qa-1 cDNA, murine NK cell cytotoxicity was inhibited by Qa-1 when Qa-1 was associated with Qdm peptide (AMAPRTLL), consistent with previous reports (Vance et al. 1998). In the presence of Yuri3 mAb, the inhibition of the target killing mediated by the NK recognition of Qa-1/Qdm complex was abrogated, indicating that CD94 on mouse NK cells is indeed involved in the recognition of Qa-1 on target cells and functions as an inhibitory receptor.

To clarify functional difference between CD94[bright] and CD94[dull] cells in Qa-1/Qdm-mediated inhibitory function, each population was isolated from C57BL/6 spleen, and cytotoxicity against CHO/Qa-1 target cells was evaluated. Importantly, Qa-1/Qdm-mediated inhibition of cytotoxicity was observed in CD94[bright] cells. The inhibition of cytotoxicity was abrogated by Yuri3 mAb while no inhibition of the cytotoxicity was observed in CD94[dull] cells. Thus, CD94[bright] but not CD94[dull] population was responsible for Qa-1 recognition.

Functional importance of mouse CD94 in self protection

We further investigated possible roles of CD94 in the self protection from NK cytotoxicity by using autologous target cells in NK assay. Activated NK cells

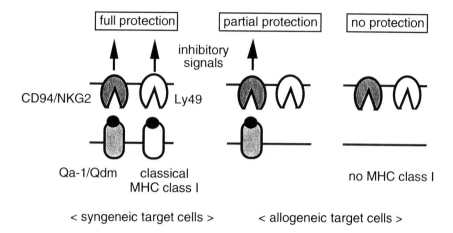

Fig. 1. Functional independence of CD94/NKG2 and Ly-49 inhibitory receptor families in target recognition by NK cells. Both CD94/NKG2 and Ly-49 family appear to be necessary for full protection from NK-mediated killing. If one of inhibitory signals mediated by these two receptor family members is absent, self tolerance of NK cells might not be guaranteed. Thus CD94/NKG2 and Ly-49 families function independently in target recognition by NK cell.

showed significant cytotoxic activity against allogeneic target cells whereas they showed very little activity against syngeneic target cells. In the presence of Yuri3 mAb, however, activated NK cells showed cytotoxicity against syngeneic target cells. This suggests that the lack of inhibitory signal through CD94/NKG2 receptor family is sufficient to induce NK cell-mediated autologous target killing. Moreover, in combination with C57BL/6-derived NK cells and BALB/c-derived target cells, NK activity against target cells was enhanced by the addition of Yuri3 mAb. C57BL/6 and BALB/c mice are allogeneic in the context of classical MHC class I but syngeneic in the context of non-classical MHC class I Qa-1[b]. Therefore, the inhibitory signal through CD94/NKG2 alone is not sufficient for the prevention of NK-mediated target killing when inhibitory signals through Ly-49 are absent. Thus, CD94 was suggested to function as an inhibitory receptor on NK cells independent of Ly-49, and both CD94/NKG2 and Ly-49 family appear to be necessary for full protection from NK-mediated killing.

Acknowledgement

We thank Ms. Y. Taguchi and Mr. T. Hosozawa for technical assistance and Ms. M. Takaku for secretarial support.

References

Braud VM, and McMichael AJ (1999) Regulation of NK cell functions through interaction of the CD94/NKG2 receptors with the nonclassical class I molecule HLA-E. Curr Topics Microbiol Immunol 244:85-95

Colonna M (1997) Specificity and function of immunoglobulin superfamily NK cell inhibitory and stimulatory receptors. Immunol Rev 155:127-133

Lanier LL (1998) Natural killer receptors. Annu Rev Immunol 16:359-393

Long EO, Wagtmann N (1997) Natural killer cell receptors. Curr Opin Immunol 9:344-350

Lopez-Botet M, Carretero M, Bellon T, Perez-Villar JJ, Llano M, Navarro F (1998) The CD94/NKG2 C-type lectin receptor complex. Curr Topics Microbiol Immunol 230:41-52

Moretta A, Biassoni R, Bottino C, Pende D, Vitale M, Poggi A, Mingari M C, Moretta L (1997) Major histocompatibility of complex class I-specific receptors on human natural killer and T lymphocytes. Immunol Rev 155:105-107

Olsson MY, Karre K, Sentman CL (1995) Altered phenotype and function of natural killer cells expressing the major histocompatibility complex receptor Ly-49 in mice transgenic for its ligand. Proc Natl Acad Sci USA 92:1649-1653

Salcedo M, Diehl A D, Olsson-Alheim M Y, Sundback J, Van Kaer L, Karre K, Ljunggren H-G (1997) Altered expression of Ly-49 inhibitory receptors on natural killer cells from MHC class I-deficient mice. J Immunol 158:3174-3180

Sivakumer PV, Gunturi A, Salcedo M, Schatzle JD, Lai WC, Kurepa Z, Pitcher L, Seaman MS, Lemonnier FA, Bennett M, Forman J, Kumar V (1999) Expression of functional CD94/NKG2A inhibitory receptors on fetal NK1.1+Ly-49- cells: a possible mechanism of tolerance during NK cell development. J Immunol 162:6976-6980

Takei F, Brennan J, Mager DL (1997) The Ly-49 family: genes, proteins and recognition of class I MHC. Immunol Rev 155:67-77

Toomey JA, Salcedo M, Cotterill LA, Millrain MM, Chrzanowska-Lightowlers Z, Lawry J, Fraser K, Gays F, Robinson JH, Shrestha S, Dyson PJ, Brooks CG (1999) Stochastic acquisition of Qa-1 receptors during the development of fetal NK cells in vitro accounts in part but not in whole for the ability of these cells to distinguish between class I-sufficient and class I-deficient targets. J Immunol 163:3176-3184

Raulet DH, Held W, Correa I, Dorfman JR, Wu M-F, Corral L (1997) Specificity, tolerance and developmental regulation of natural killer cells defined by expression of class I-specific Ly49 receptors. Immunol Rev 155:41-52

Vance RE, Kraft JR, Altman JD, Jensen PE, Raulet DH (1998) Mouse CD94/NKG2A is a natural killer cell receptor for the nonclassical major histocompatibility complex (MHC) class I molecule Qa-1^b. J Exp Med 188:1841-1848

Yokoyama WM (1993) The Ly-49 and NKR-P1 gene families encoding lectin-like receptors on natural killer cells: The NK gene complex. Annu Rev Immunol 11:613-635

12
Interaction of human cytomegalovirus glycoproteins with immunoreceptors

David Cosman, Jan Chalupny, Mei-Ling Hsu, Claire Sutherland, Jürgen Müllberg, Marek Kubin, Neil Fanger, and Luis Borges

Immunex Corporation, 51 University St., Seattle, WA 98110, USA

Summary. The use of glycoproteins encoded by human cytomegalovirus as probes to isolate their cellular counterstructures has resulted in the discovery of novel immunoreceptors. The HCMV-encoded MHC class I homolog, UL18, binds to LIR-1/ILT2, an inhibitory signaling receptor for cellular MHC class I antigens with a broad distribution on leukocytes, including some NK cells. Although UL18 has been proposed to act as an inhibitor of NK cytotoxicity, this remains controversial. Another HCMV-encoded glycoprotein, UL16, binds to members of a novel non-classical MHC class I-related family, the ULBPs, as well as to MICB, a known non-classical MHC class I antigen. The MICs and ULBPs are ligands for the activating receptor, NKG2D/DAP10, expressed by NK and other immune effector cells. Ligation of NKG2D/DAP10 by ULBPs or MICs on a target cell can overcome an inhibitory signal mediated by NK recognition of MHC class I antigens and allow NK cytotoxicity. UL16 masking of ULBP or MIC recognition may represent a mechanism of immune evasion by CMV.

Key words. UL18, UL16, NKG2D, NK cells, LIR-1/ILT2

Introduction

During the co-evolution of a virus with its host, the virus must evolve mechanisms for evading or counteracting the host's immune response. Research over the past few years has revealed an impressive variety of tricks that viruses use for this purpose. These include virus-encoded cytokines, chemokines, soluble cytokine receptors, chemokine binding proteins, chemokine receptors, complement inhibitors, intracellular inhibitors of cytokine signaling, and caspase inhibitors (Spriggs 1996). Given that the large DNA viruses, such as the poxvirus and herpesvirus families, encode many genes of unknown function, there is no reason to suspect that we have discovered all the mechanisms by which these viruses can subvert the immune response.

Our group has focused on human cytomegalovirus (HCMV) as a potentially rich source of immunomodulatory genes; study of which may lead to the discovery of new immune system components and provide new insights into control of immune responses. HCMV co-exists with its host in an apparent state of latency, but is able to undergo periodic inapparent reactivation with low levels of virus shedding despite the presence of a functional immune response. This demonstrates that HCMV has evolved effective mechanisms to deflect this response. In immunocompromised individuals, the virus is released from the restraints normally imposed and creates considerable damage to multiple organ systems (Britt and Alford 1996). The genome of HCMV is large, and contains many open reading frames predicted to encode glycoproteins (Chee et al. 1990). Some of these are known to be non-essential for replication in tissue culture, and are therefore candidate immunomodulators. Our approach has been to try and identify cellular proteins with which these viral proteins interact.

Only within the past few years has there been a recognition that cellular responses are controlled not only by signals received and transmitted through cell surface stimulatory or activating receptors, but that inhibitory signalling receptors are also important. Furthermore, the inhibitory receptors often have the same ligand specificity as the activating receptor, thus providing a fine balance in controlling cellular responses (Bakker et al. 2000; Long 1999; Ravetch and Lanier 2000). This phenomenon was initially recognised in two systems, the discovery of activating and inhibitory Fc receptors on B cells, and the control of NK cytotoxicity where NK inhibitory receptors for MHC class I antigens on target cells are balanced by activating receptors for the same ligands (Ravetch and Lanier 2000). More recently, multiple families of inhibitory and activating receptors have been identified on all classes of hematopoietic cells, as well as other cell types, underscoring the generality of this type of regulation (Colonna et al. 2000). Given the importance of these types of immunoreceptors, it is perhaps not so suprising that viruses have targeted them. In this review, I will describe our findings that HCMV encodes proteins that may disrupt the functions of both activating and inhibitory receptors.

UL18

UL18 is an HCMV open reading frame that encodes a protein homologous to MHC class I antigens (Beck and Barrell 1988). It is known to bind peptides and to associate with β-2 microglobulin (Browne et al. 1990; Fahnestock et al. 1995). Although UL18 was initially hypothesized to be responsible for down-regulation by HCMV of the expression of host MHC class I antigens by competing for β-2 microglobulin, this was later shown not to be true (Browne et al. 1992). A later hypothesis proposed that UL18 might engage NK cell killer inhibitory receptors (KIR) for host MHC class I antigens and protect infected cells from NK cytotoxicity (Fahnestock et al. 1995). In our own studies, we searched directly for the UL18 counterstructure by constructing a UL18-Fc fusion protein using the extracellular domain of UL18. This protein bound strongly to several B and myeloid cell lines, and expression cloning of the UL18 counterstructure revealed a novel immunoglobulin superfamily receptor that we called LIR-1 (**L**eukocyte **I**mmunoglobulin-like **R**eceptor) (Cosman et al. 1997). The same molecule was independently cloned by a different route and called ILT2 (Samaridis and Colonna 1997).

LIR-1 contained four immunoglobulin-like domains in the extracellular region and four ITIM-like (immunoreceptor tyrosine-based inhibitory motifs) sequences in the cytoplasmic domain. Interestingly, its sequence showed significant homology to the NK Ig-like KIRs, which also contain ITIMs. However, unlike KIRs, which are expressed on NK cells and some T cells, LIR-1 is expressed on all B and myeloid cells, as well as a variable subset of NK and T cells (Cosman et al. 1997). Like KIRs, LIR-1 was shown to recognize cellular MHC class I antigens, but with distinct specificity. Whereas individual KIRs recognize subsets of HLA-A, B, or C alleles, LIR-1 interacts with most classical MHC class I antigens and also with HLA-G (Colonna et al. 1997; Colonna et al. 1998; Cosman et al. 1997; Fanger et al. 1998). The finding of receptors for MHC class I on B cells and myeloid cells implied a role for self recognition by these cells that had not been previously recognized. The biological implications of this are yet to be established. Interestingly, UL18 bound to LIR-1 much more strongly than did host MHC class I antigens. This enhanced binding affinity allowed the elucidation of the mode by which UL18 (and by implication host MHC class I antigens) bind to LIR-1 (Chapman et al. 1999). The N-terminal Ig-like domain of LIR-1 contacts the $\alpha 3$ domain of UL18, in contrast to the KIRs, which bind to the $\alpha 1$ domain of MHC class I antigens (Fan et al. 1997).

Like the KIRs, the LIRs are a multi-gene family mapping to chromosome 19q13.4. Eleven LIR/ILT loci have been recognized (Wende et al. 2000), and the family of encoded proteins contains members with two or four Ig-like domains, and sequences characteristic of both inhibitory and stimulatory receptors (Borges et al. 1997; Colonna et al. 2000). Only one other LIR, LIR-2 (ILT4), has been shown to recognize MHC class I antigens (Borges et al. 1997; Fanger et al. 1998; Colonna et al. 1998). Interestingly, LIR-2 does not interact with UL18, and also differs from LIR-1 in its expression pattern, being restricted to myeloid cells such as macrophages, neutrophils and dendritic cells (Cosman et al. 1997; Fanger et al. 1998).

The function of UL18 remains controversial. An initial report claimed that expression of UL18 in MHC class I negative cells could protect against NK cytotoxicity, but also claimed that this effect was mediated through CD94 on the NK cells (Reyburn et al. 1997). As UL18 does not detectably bind to CD94, but does bind strongly to LIR-1, this seems unlikely (Cosman et al. 1999). Subsequent studies from other groups failed to show UL18-mediated protection against NK cytotoxicity, and found that fibroblast cells infected by a UL18 deficient HCMV strain were actually more resistant to NK cytotoxicity than cells infected by wild type HCMV (Leong et al. 1998). It has also been found that resistance of CMV-infected cells to NK cytotoxicity in vitro correlated better with expression of adhesion molecules, such as LFA-3, on the target cell rather than the levels of MHC class I (Fletcher et al. 1998).

How UL18 functions in vivo is unknown. As discussed, it might target LIR-1 on cells of the monocytic lineage, either to inhibit the production of anti-viral cytokines, or to influence the differentiation state of the cell (Cosman et al. 1999).

UL16

Unlike UL18, the HCMV ORF, UL16, has no detectable homology to other proteins. UL16 is expressed in HCMV-infected cells but is not detected on virions, and is non-essential for HCMV growth in vitro. The sequence of UL16 predicts a type 1 membrane glycoprotein (Kaye et al. 1992). In order to identify its counterstructure, we constructed a UL16-Fc fusion protein and used it in flow cytometry experiments. Human T (HSB-2) and B (Namalwa) cell lines showed specific binding of UL16-Fc. As cDNA expression libraries for these cell lines were available, we screened them for UL16-Fc binding. Two distinct cDNAs were isolated. One, from HSB2, corresponded to an allele of MICB, a polymorphic, non-classical, MHC class I gene (Bahram et al. 1994; Leelayuwat et al. 1994) the other, from Namalwa, was a novel cDNA that we called ULBP1 (**UL16 B**inding **Pro**tein). Subsequently, two closely related cDNAs were identified in the GenBank EST database, and named ULBP2 and ULBP3. The ULBP molecules are also distantly related to MHC class I antigens, but contain only $\alpha 1$ and $\alpha 2$ domains, no $\alpha 3$ domain, and are membrane anchored via GPI linkages. Unlike most MHC class 1 genes, including the MICs, the ULBP genes are clustered on chromosome 6q25, outside the major histocompatibilty complex (Fig. 1).

Interestingly, UL16 binds to MICB, but not the closely-related MICA protein, and to ULBP1 and ULBP2 but not the closely related ULBP3 protein. This could be interpreted as supporting distinct biological roles for different members of the MIC and ULBP families, or could reflect polymorphism in either the cellular or viral genes. The MICs are known to be very polymorphic, and we have only investigated UL16 binding to two alleles of MICB and one of MICA. Polymorphism in the ULBP genes or between UL16 genes from different viral strains has not yet been analyzed.

In order to examine the function of the ULBP molecules, we constructed soluble Fc and leucine zipper fusion proteins. These proteins were used in flow cytometry experiments and found to bind reproducibly to human NK cells. Binding was up-regulated by IL-15 pretreatment of the NK cells. We found that the soluble ULBP-LZ fusion proteins stimulated the production of a number of

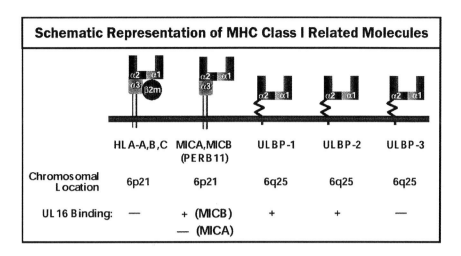

Fig. 1. Schematic representation of ULBP, MIC and classical MHC class I molecules. Chomosomal location of the genes and UL16 binding are indicated.

cytokines and chemokines from the NK cells, and that in all cases, co-stimulation of the NK cells with IL-12 enhanced cytokine and chemokine production. The degree of synergy between IL-12 and the ULBPs varied greatly between different cytokines and chemokines.

Recently, the receptor for the MICs was shown to be a heterocomplex between NKG2D, a member of the C-type lectin family and DAP10, a small transmembrane adaptor protein that is involved in signal transduction (Bauer et al. 1999; Houchins et al. 1990; Wu et al. 1999). Given that some ULBPs and MICs share binding to UL16, and that NKG2D/DAP10 is expressed on NK cells, we tested the binding of the ULBPs to recombinantly expressed NKG2D/DAP10 and found that all three ULBPs bound strongly. While this does not prove that NKG2D/DAP10 is the entire ULBP receptor on NK cells or other cells, we were able to show that a polyclonal antiserum to NKG2D was able to block ULBP binding to NK cells, suggesting that NKG2D is at least a component of the receptor for the ULBPs.

An important function of NKG2D/DAP10 is the activation of NK cytotoxicity, even against MHC class I positive target cells (Bauer et al. 1999). We asked whether the ULBPs could induce cytotoxicity in the same way. Daudi cells, which are MHC class I negative due to a mutation in β-2 microglobulin, are good targets for NK cells. Transduction of these cells by β-2 microglobulin resulted in re-expression of high levels of MHC class I and relative resistance to NK cytotoxicity. When individual ULBP or MIC cDNAs were transduced into the MHC class I-expressing Daudi cells, in each case the cells were rendered much more susceptible to NK cytotoxicity. A Fab fragment of a monoclonal antibody directed against ULBP1 completely blocked cytotoxicity directed against the ULBP1-expressing Daudi targets. This result has implications for the ability of the innate immune system to function against MHC class I positive tumor cells or pathogen-infected cells in which MHC class I levels are not perturbed. ULBPs and MICs may be important sentinel molecules that are recognized by NKG2D-expressing cells (NK cells, γδ T cells, CD8[+] T cells and activated macrophages) to trigger cytotoxicity as well as the release of cytokines and chemokines that can recruit and activate other cells of the innate and adaptive immune systems.

The finding that the ULBPs and MICs can mediate these functions immediately suggests a model for how UL16 might protect an infected cell against immune recognition. UL16, in soluble form, can block the binding and biological activity of soluble ULBPs on NK cells. It is plausible that UL16, expressed in a virus-infected cell, might mask the recognition of ULBP1, ULBP2 or MICB by NKG2D-epressing cells, thereby preventing cytotoxicity (Fig. 2). UL16 and the ULBPs might interact on the cell surface, or they might interact intracellularly to prevent ULBP surface expression. This model remains to be tested.

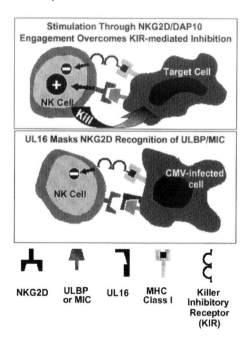

Fig. 2. A model for UL16-mediated immune evasion. Top panel: recognition of NKG2D ligands (ULBPs or MICs) on target cells allows NK cytotoxicity--even against MHC class I expressing targets. Bottom panel: Recognition of NKG2D ligands is masked by UL16 expression in the target cell, which prevents NK cytotoxicity.

Our studies have shown that HCMV glycoproteins have targeted both inhibitory and stimulatory immunoreceptors. While we currently lack a detailed understanding of the mechanisms by which these interactions benefit the virus, it is fair to say that the study of these mechanisms should reveal new insight into immune regulation.

References

Bakker AB, Wu J, Phillips JH, Lanier LL (2000) NK cell activation: distinct stimulatory pathways counterbalancing inhibitory signals. Hum Immunol 61:18-27

Bauer S, Groh V, Wu J, Steinle A, Phillips JH, Lanier LL, Spies T (1999) Activation of NK cells and T cells by NKG2D, a receptor for stress- inducible MICA. Science 285:727-729

Beck S, Barrell BG (1988) Human cytomegalovirus encodes a glycoprotein homologous to MHC class-I antigens. Nature 331:269-272

Britt W, Alford C (1996) Cytomegalovirus. In: Field BN, Knipe DM, Howley PM (Eds) Field's Virology. Lippincott-Raven, Philadelphia, pp 2493-2523

Borges L, Hsu ML, Fanger N, Kubin M, Cosman D (1997) A family of human lymphoid and myeloid Ig-like receptors, some of which bind to MHC class I molecules. J Immunol 159:5192-5196

Browne H, Churcher M, Minson T (1992) Construction and characterization of a human cytomegalovirus mutant with the UL18 (class I homolog) gene deleted. J Virol 66:6784-6787

Browne H, Smith G, Beck S, Minson T (1990) A complex between the MHC class I homologue encoded by human cytomegalovirus and beta 2 microglobulin. Nature 347:770-772

Chapman TL, Heikeman AP, Bjorkman PJ (1999) The inhibitory receptor LIR-1 uses a common binding interaction to recognize class I MHC molecules and the viral homolog UL18. Immunity 11:603-613

Chee MS, Bankier AT, Beck S, Bohni R, Brown CM, Cerny R, Horsnell T, Hutchison CAD, Kouzarides T, Martignetti JA, et al. (1990) Analysis of the protein-coding content of the sequence of human cytomegalovirus strain AD169. Curr Top Microbiol Immunol 154:125-169

Colonna M, Nakajima H, Cella M (2000) A family of inhibitory and activating Ig-like receptors that modulate function of lymphoid and myeloid cells. Semin Immunol 12:121-127

Colonna M, Navarro F, Bellon T, Llano M, Garcia P, Samaridis J, Angman L, Cella M, Lopez-Botet M (1997) A common inhibitory receptor for major histocompatibility complex class I molecules on human lymphoid and myelomonocytic cells. J Exp Med 186:1809-1818

Colonna M, Samaridis J, Cella M, Angman L, Allen RL, O'Callaghan CA, Dunbar R, Ogg GS, Cerundolo V, Rolink A (1998) Human myelomonocytic cells express an inhibitory receptor for classical and nonclassical MHC class I molecules. J Immunol 160:3096-3100

Cosman D, Fanger N, Borges L (1999) Human cytomegalovirus, MHC class I and inhibitory signalling receptors: more questions than answers. Immunol Rev 168:177-185

Cosman D, Fanger N, Borges L, Kubin M, Chin W, Peterson L, Hsu ML (1997) A novel immunoglobulin superfamily receptor for cellular and viral MHC class I molecules. Immunity 7:273-282

Fahnestock ML, Johnson JL, Feldman RM, Neveu JM, Lane WS, Bjorkman PJ (1995) The MHC class I homolog encoded by human cytomegalovirus binds endogenous peptides. Immunity 3:583-590

Fan QR, Mosyak L, Winter CC, Wagtmann N, Long EO, Wiley DC (1997) Structure of the inhibitory receptor for human natural killer cells resembles haematopoietic receptors. Nature 389:96-100

Fanger NA, Cosman D, Peterson L, Braddy SC, Maliszewski CR, Borges L (1998) The MHC class I binding proteins LIR-1 and LIR-2 inhibit Fc receptor- mediated signaling in monocytes. Eur J Immunol 28:3423-3434

Fletcher JM, Prentice HG, Grundy JE (1998) Natural killer cell lysis of cytomegalovirus (CMV)-infected cells correlates with virally induced changes in cell surface lymphocyte function-associated antigen-3 (LFA-3) expression and not with the CMV- induced down-regulation of cell surface class I HLA. J Immunol 161:2365-2374

Kaye J, Browne H, Stoffel M, Minson T (1992) The UL16 gene of human cytomegalovirus encodes a glycoprotein that is dispensable for growth in vitro. J Virol 66:6609-6615

Leong CC, Chapman TL, Bjorkman PJ, Formankova D, Mocarski ES, Phillips JH, Lanier LL (1998) Modulation of natural killer cell cytotoxicity in human cytomegalovirus infection: the role of endogenous class I major histocompatibility complex and a viral class I homolog. J Exp Med 187:1681-1687

Long EO (1999) Regulation of immune responses through inhibitory receptors. Annu Rev Immunol 17:875-904

Ravetch JV, Lanier LL (2000) Immune inhibitory receptors. Science 290:84-89

Reyburn HT, Mandelboim O, Vales-Gomez M, Davis DM, Pazmany L, Strominger JL (1997) The class I MHC homologue of human cytomegalovirus inhibits attack by natural killer cells. Nature 386:514-517

Samaridis J, Colonna M (1997) Cloning of novel immunoglobulin superfamily receptors expressed on human myeloid and lymphoid cells: Structural evidence for new stimulatory and inhibitory pathways. Eur J Immunol 27:660-665

Spriggs MK (1996) One step ahead of the game: viral immunomodulatory molecules. Annu Rev Immunol 14:101-130

Wende H, Volz A, Ziegler A (2000) Extensive gene duplications and a large inversion characterize the human leukocyte receptor cluster. Immunogenetics 51:703-713

13
Function of gp49A in mast cell activation

Masao Ono, Kwang Ho Lee, and Toshiyuki Takai

Department of Experimental Immunology and CREST Program of JST, Institute of
Development, Aging and Cancer, Tohoku University, Seiryo 4-1, Sendai 980-8575, Japan

Summary. Type-I transmembrane glycoprotein, gp49A, is a murine C2-type
immunoglobulin-like receptor expressed on the cells involved in natural immunity
in parallel with its closely related isotype, gp49B. It has been demonstrated that
gp49B contains two immunoreceptor tyrosine-based inhibitory motifs (ITIMs) in
its cytoplasmic region, and thereby potentially act as an inhibitory receptor on
mast cells and NK cells by associating tyrosine phosphatase activity to
phosphorylated ITIMs. On the other hand, as gp49A does not harbor any specific
motif for signal transduction, its physiological role has not been determined. Here,
we present some novel evidences indicating potential function of gp49A in mast
cell activation. Crosslinking of gp49A on rat mast cell line RBL-2H3 evoked
cytoplasmic calcium mobilization, prostaglandin D_2 (PGD_2) release, and
transcription of IL-3 and IL-4 genes, but did not elicit exocytosis. Furthermore,
activation of gp49-mediated signal required constitutive receptor dimerization and
inducible incorporation into glycolipid-enriched membrane fraction (GEM).

Key words. gp49A, Mast cells, Activation, GEM, Microdomain

Introduction

Investigations on the cell-associated antigen receptors such as TCR and surface
immunoglobulin receptor on B cell (sIg) have delineated signaling mechanisms
for cellular activation resulting in antigen-specific (acquired) immune response. In
the meantime, a number of immunoglobulin (Ig)-like receptors have been
identified and demarcated the superfamily of Ig-like receptors (Ig-superfamily). In
the last decade, the growing number of Ig-like receptors put forward these two
important notions that;
- several subfamilies among members of Ig-superfamily were defined by the
  criteria of extensive similarity of ectodomain and clustered gene location on a
  chromosome, and that

- most of these subfamilies consist of inhibitory and stimulatory isotypes of receptors.

For examples, human killer cell Ig-like receptors (KIR) (Colonna and Samaridis 1995; Wagtmann et al. 1995; Wilson et al. 2000), human Ig-like transcript/leukocyte Ig-like receptor/myeloid Ig-like receptor (ILT/LIR/MIR) (Samaridis and Colonna 1997; Cosman et al. 1997; Wagtmann et al. 1997; Arm et al. 1997; Wilson et al. 2000), signal induction receptor proteins (SIRP) (Ohnishi et al. 1996; Fujioka et al. 1996; Kharitonenkov et al. 1997), and paired Ig-like receptors (PIR) (Hayami et al. 1997; Kubagawa et al. 1997; Yamashita et al. 1998) are of these notions. Presently, two distinct features can be attributed to inhibitory and stimulatory function of Ig-like receptors. The inhibitory Ig-like receptors commonly harbor particular amino acid motifs, termed as immunoreceptor tyrosine-based inhibitory motif (ITIM) in their cytoplasmic region (CP). It has been demonstrated that ITIM plays the mandatory and initial role in inhibitory signal transduction by docking SH2-containing tyrosine phsosphatase-1 (SHP-1) upon tyrosine-phosphorylation of ITIM (Ono et al. 1997). On the other hand, the stimulatory Ig-like receptors in themselves lack know motif sequence in CP, but commonly have single basic amino acid residue in their transmembrane regions (TM). It has been revealed that this basic amino acid residue in TM is critical for associating with such signaling subunits as FcεRIγ or DAP12/KARAP, either of which contains immunoreceptor tyrosine-based activation motif (ITAM) to activate signal transduction (Maeda et al. 1998; Ono et al. 1999; Nakajima et al. 1999; Lanier et al. 1998; Tomasello et al. 1998). Since these Ig-like receptor subfamilies are mainly expressed on the cell types for innate (natural) immune system, the great interest has been posed in the role of their positive and negative regulations in innate immune responses.

Murine gp49s, type-I transmembrane glycoproteins, are composed mainly of two structurally distinct isoforms, gp49A and gp49B (Castells et al. 1994). Recent studies including the analysis using gp49B-deficient mice have clearly revealed that both isoforms are expressed on IL-2 dependent mouse NK cells, IL-3 dependent mouse bone marrow derived mast cells, peritoneal macrophages and several mouse mast cell lines, although the gp49A expression on NK cells are found very few (Rojo et al. 2000; our unpublished observation). The two ectodomains, TM and membrane-proximal CP of gp49 are extremely conserved between two isotypes by 89% amino acid identity. Their diversity mainly exists in their CP. The gp49B harbors two ITIM sequences in its CP, but no ITIM sequence exists in the gp49A. Functionally, gp49B was found to exert inhibitory effect on mast cell exocytosis and NK cell activation under the experimental condition (Katz et al. 1996; Rojo et al. 1997). The entire nucleotide sequences of gp49A and gp49B genes revealed their extensive similarity over exons and introns by more than 94% (Kuroiwa et al. 1998; McCormick et al. 1999). However two cysteine residues are added in exon 5 of gp49A which corresponds to the residues upstream of the first ITIM of gp49B, resulting in lacking ITIM in gp49A while intact sequence for two ITIMs remains untranslated in 3' flanking region of gp49A (Katz et al. 1996). The mutual conservation of gp49 genes suggests that gp49 genes were generated by gene duplication of a gp49-prototype gene, and that gp49A is a loss-of-function descendent of inhibitory prototype receptor. In contrast, another subfamilies of Ig-like receptors such as KIR, ILT/LIR/MIR and PIR don't follow this insight. In these subfamilies, the portion including TM and CP of non-inhibitory isoforms is quite different in germline gene level from that of inhibitory isoforms, so that functional diversity of two isotypes can not be interpreted by the gene duplication of single prototype gene as simply as in the gp49 case (Yamashita et al. 1998; Wilson et al. 2000). The genes of gp49A and gp49B are located in chromosome 10 B4 region, and exhibit no polymorphism among the mouse strains examined (Kuroiwa et al. 1998). This chromosonal region is not syntenic and neighboring to the locus for NK gene complex (NKC) and leukocyte receptor complex (LRC) in human and mouse. These facts and insights have posed the speculation that gp49A may be a kind of decoy receptor for gp49B, and that gp49 have evolved in the context of the biological significance different from that for another subfamilies of Ig-like receptors such as KIR, ILT/LIR/MIR and PIR.

gp49A is potentially effective for mast cell activation

Despite the fact that gp49A lacks any consensus amino acid motif to be involved in signal transduction and subunit association, we recently suggested that aggregation of gp49A could trigger signal transduction in mast cells (Lee et al. 2000). Indeed in our experiments, aggregation of chimeric receptor composed of intact pre-TM, TM, CP region of gp49A and the ectodomains of mouse FcγRII (FcR-gp49A) elicited such activation traits as cytoplasmic calcium mobilization, some cytokine expressions and proinflammatory eicosanoid PGD_2 synthesis in rat mast cell line RBL-2H3, whereas exocytosis is not inducible under the same condition. Cytoplasmic calcium mobilization can be observed even in the absence of the extracellular calcium, suggesting involvement of phospholipase C activation in the downstream of gp49A-mediated signal transduction. As compared with the nature of FcεRI-mediated signal transduction, gp49A-mediated signal evokes much lower level of calcium mobilization in the delayed manner, yet it evokes comparable level of cytokine gene activation and PGD_2 synthesis in RBL-2H3 cell. These findings may suggest that gp49A favors triggering of the signal cascade linked to gene activation of cytokines and eicosaniod synthesis in mast cells, and that the signaling pathway from calcium mobilization to exocytotic response is defective or insufficient upon aggregation of gp49A.

How does gp49A mediate signal?

A large number of receptors for cellular activation exist in the form of multimeric protein complex, in which a few components act as intracellular mediator for signal transduction. In this respect, we examined the association of another receptor components with FcR-gp49A in RBL-2H3 cell. Even using such low-stringent detergents as digitonin and Brij 97 for extraction of membrane proteins, we couldn't detect another protein species associated with FcR-gp49A though a possibility still remains that association of another protein is so weak as to be undetectable. In the meantime, these experiments brought us an important notion that FcR-gp49A exist as homodimer on RBL-2H3 cell in the manner sensitive for reducing reagent. The result from the experiments using mutant FcR-gp49A, in which single cysteine residue (Cys_{226}) in pre-TM was replaced with phenylalanine, has clearly revealed that Cys_{226} is critical for homophilic dimerization of FcR-gp49A. Importantly, this mutation on Cys_{226} also removed the capacity to trigger signal transduction from FcR-gp49A, yet its membrane expression was not affected. These findings indicate that steady dimerization of gp49A is prerequisite to its function for mast cell activation. Previous study demonstrated that serine residue on gp49B in bone marrow-derived mast cell could be phosphorylated in response to the stimulation of IgE plus antigen, phorbol ester or calcium ionophor A23187 (Katz et al. 1989). Since the potential residue (Ser_{273}) of protein kinase C phosphorylation of gp49B is conserved in gp49A, involvement of Ser_{273} in gp49A-mediated signal transduction could have been postulated. But presently, FcR-gp49A with single mutation of Ser_{273} was found to trigger the signal transduction to the same extent as intact FcR-gp49A (our unpublished data), thereby it is unlikely that phosphorylation of Ser_{273} is involved in gp49A function for mast cell activation.

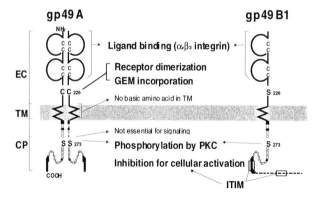

Fig. 1. Functioanl features of gp49A and gp49B attributable to their structure. In CP, identical and different amino acid residues are indicated by open and closed circles, respectively. EC; extracellular region, PKC; protein kinase C.

Particular membrane microdomain is involved in gp49A function

The FcR-gp49A does not associate with FcεRIγ subunit and DAP-12, the latter of which has recently been identified as a novel signaling subunit containing ITAM in association with some receptors (Lanier et al. 1998B; Smith et al. 1998; Lanier et al. 1998A). In addition, FcR-gp49A aggregation never induces tyrosine-phosphorylation of FcεRIγ and DAP-12, though RBL-2H3 has been found to lack DAP-12 expression (our unpublished data). Apparently, we could postulate a different mechanism independent of ITAM for gp49A-mediated signal transduction. For the function of a number of the membrane-integrated immune receptors including B cell antigen receptor, T cell receptor and Fc receptors, particular membrane microdomain has been shown to play an important role. This plasma membrane structure is otherwise denoted as glycolipid-enriched membrane domain (GEM) or membrane raft. Biochemically, GEM can be distinguished from the rest of the plasma membrane by their relative insolubility in detergents and low buoyant density (Brown and Rose 1992). Structurally, GEM accumulates the molecules of importance in signal transduction on inner layer such as tyrosine kinases and various adapter proteins (Horejsi et al. 1999). Therefore functionally, GEM is believed to be the effective site to transmit stimuli to intracellular signaling cascade. As an another function of GEM, the outer layer of GEM serves as the scaffolding site for glycosyl-phosphatidylinositol (GPI)-anchored membrane proteins (Brown and Rose 1992; Deckert et al. 1996). Crosslinking of GPI-anchored receptors may alter the GEM structure. Thereby GEM function is implicated in signaling mechanism of such GPI-anchored receptors as Thy-1 and human FcγRIIIb (Deckert et al. 1996). We have shown that GEM is also implicated in the gp49A-mediated signal transduction. As aggregated with anti-FcR antibody (2.4G2) to activate signal transduction in RBL-2H3, intact FcR-gp49A was found in the GEM isolated as Triton X-100-insoluble fraction followed by density-gradient ultracentrifugation, whereas Cys_{226}-mutant FcR-gp49A was not. If not aggregated, intact FcR-gp49A could not be detected in GEM fraction. Accordingly, the receptor dimerization and massive aggregation are necessary for incorporation of FcR-gp49A into GEM. Furthermore, we found that FcR-gp49A-mediated signaling event in RBL-2H3 cell was significantly attenuated by the treatment with methyl-β-cyclodextrin (MBCD), which disrupts GEM structure by extracting cholesterol from plasma membrane. Taken together, it is suggested that the function of gp49A depends on the GEM structure of mast cells.

Concluding remarks

Of note, our current findings uncovered a role of gp49A function in mast cell activation, and a new way of interpretation for triggering signal transduction by the Ig-like receptor lacking any predictable motifs for cellular activation. Like another subfamilies of Ig-like receptor such as KIR, ILT/LIR/MIR and PIR, the gp49 subfamily also seems to be involved in the mechanisms to regulate immune response by virtue of its positive and negative effects on immune effector functions. Apparently, before concluding gp49 functions in immune system as mentioned above, more detailed analyses should be needed especially in these four points of view.

- Can gp49A crosslinking activate primary mast cells?
- What is the ligand for gp49A?
- What is the immunological situation to activate gp49A and B?
- Can gp49B inhibit gp49A? It's not obvious, because it remains unclear that gp49A activates lyn kinase, which was thought to be central to phosphorylate ITIM.

Furthermore, it should be of physiological importance to address two questions as to how the dimer gp49A moves to GEM (in another words, what the significance of gp49A dimerization in its GEM incorporation is) and how the signal can be transduced downstream of GEM independently of ITAM.

Acknowledgments

We would like to thank M. Inui for his helpful discussion and efforts on searching necessary information in manuscript preparation.

References

Arm JP, Nwankwo C, Austen KF (1997) Molecular identification of a novel family of human Ig superfamily members that possess immunoreceptor tyrosine-based inhibition motifs and homology to the mouse gp49B1 inhibitory receptor. J Immunol 159:2342-2349

Brown DA, Rose JK (1992) Sorting of GPI-anchored proteins to glycolipid-enriched membrane subdomains during transport to the apical cell surface. Cell 68:533-544

Castells MC, Wu X, Arm JP, Austen KF, Katz HR (1994) Cloning of the gp49B gene of the immunoglobulin superfamily and demonstration that one of its two products is an early-expressed mast cell surface protein originally described as gp49. J Biol Chem 269 8393-8401

Colonna M, Samaridis J (1995) Cloning of immunoglobulin-superfamily members associated with HLA-C and HLA-B recognition by human natural killer cells. Science 268:405-408

Cosman D, Fanger N, Borges L, Kubin M, Chin W, Peterson L, Hsu ML (1997) A novel immunoglobulin superfamily receptor for cellular and viral MHC class I molecules. Immunity 7:273-282

Deckert M, Ticchioni M, Bernard A (1996) Endocytosis of GPI-anchored proteins in human lymphocytes: role of glycolipid-based domains, actin cytoskeleton, and protein kinases. J Cell Biol 133:791-799

Fujioka Y, Matozaki T, Noguchi T, Iwamatsu A, Yamao T, Takahashi N, Tsuda M, Takada T, Kasuga M (1996) A novel membrane glycoprotein, SHPS-1, that binds the SH2-domain-containing protein tyrosine phosphatase SHP-2 in response to mitogens and cell adhesion. Mol Cell Biol 16:6887-6899

Hayami K, Fukuta D, Nishikawa Y, Yamashita Y, Inui M, Ohyama Y, Hikida M, Ohmori H, Takai T (1997) Molecular cloning of a novel murine cell-surface glycoprotein homologous to killer cell inhibitory receptors. J Biol Chem 272:7320-7327

Horejsi V, Drbal K, Cebecauer M, Cerny J, Brdicka T, Angelisova P, Stockinger H (1999) GPI-microdomains: a role in signalling via immunoreceptors. Immunol Today 20:356-361

Katz HR, Benson AC, Austen KF (1989) Activation- and phorbol ester-stimulated phosphorylation of a plasma membrane glycoprotein antigen expressed on mouse IL-3-dependent mast cells and serosal mast cells. J Immunol 142:919-926

Katz HR, Vivier E, Castells MC, McCormick MJ, Chambers JM, Austen KF (1996) Mouse mast cell gp49B1 contains two immunoreceptor tyrosine-based inhibition motifs and suppresses mast cell activation when coligated with the high-affinity Fc receptor for IgE. Proc Natl Acad Sci USA 93:10809-10814

Kharitonenkov A, Chen Z, Sures I, Wang H, Schilling J, Ullrich A (1997) A family of proteins that inhibit signalling through tyrosine kinase receptors. Nature 386:181-186

Kubagawa H, Burrows PD, Cooper MD (1997) A novel pair of immunoglobulin-like receptors expressed by B cells and myeloid cells. Proc Natl Acad Sci USA 94:5261-5266

Kuroiwa A, Yamashita Y, Inui M, Yuasa T, Ono M, Nagabukuro A, Matsuda Y, Takai T (1998) Association of tyrosine phosphatases SHP-1 and SHP-2, inositol 5-phosphatase SHIP with gp49B1, and chromosomal assignment of the gene. J Biol Chem 273:1070-1074

Lanier LL, Corliss BC, Wu J, Leong C, Phillips JH (1998A) Immunoreceptor DAP12 bearing a tyrosine-based activation motif is involved in activating NK cells. Nature 391: 703-707

Lanier LL, Corliss B, Wu J, Phillips JH (1998B) Association of DAP12 with activating CD94/NKG2C NK cell receptors. Immunity 8:693-701

Lee KH, Ono M, Inui M, Yuasa T, Takai T (2000) Stimulatory function of gp49A, a murine Ig-like receptor, in rat basophilic leukemia cells. J Immunol 165, 4970-4977

Maeda A, Kurosaki M, Kurosaki T (1998) Paired immunoglobulin-like receptor (PIR)-A is involved in activating mast cells through its association with Fc receptor γ-chain. J Exp Med 188:991-995

McCormick MJ, Castells MC, Austen KF, Katz HR (1999) The gp49A gene has extensive sequence conservation with the gp49B gene and provides gp49A protein, a unique member of a large family of activating and inhibitory receptors of the immunoglobulin superfamily. Immunogenetics 50:286-294

Nakajima H, Samaridis J, Angman L, Colonna M (1999) Human myeloid cells express an activating ILT receptor (ILT1) that associates with Fc receptor γ-chain. J Immunol 162:5-8

Ohnishi H, Kubota M, Ohtake A, Sato K, Sano Si (1996) Activation of protein-tyrosine phosphatase SH-PTP2 by a tyrosine-based activation motif of a novel brain molecule. J Biol Chem 271:25569-25574

Ono M, Okada H, Bolland S, Yanagi S, Kurosaki T, Ravetch JV (1997) Deletion of SHIP or SHP-1 reveals two distinct pathways for inhibitory signaling. Cell 90:293-301

Ono M, Yuasa T, Ra C, Takai T (1999) Stimulatory function of paired immunoglobulin-like receptor-A in mast cell line by associating with subunits common to Fc receptors. J Biol Chem 274:30288-30296

Rojo S, Burshtyn DN, Long EO, Wagtmann N (1997) Type I transmembrane receptor with inhibitory function in mouse mast cells and NK cells. J Immunol 158:9-12

Rojo S, Stebbins CC, Peterson ME, Dombrowicz D, Wagtmann N, Long EO (2000) Natural killer cells and mast cells from gp49B null mutant mice are functional. Mol Cell Biol 20:7178-7182

Samaridis J, Colonna M (1997) Cloning of novel immunoglobulin superfamily receptors expressed on human myeloid and lymphoid cells: structural evidence for new stimulatory and inhibitory pathways. Eur J Immunol 27, 660-665

Smith KM, Wu J, Bakker AB, Phillips JH, Lanier LL (1998) Ly-49D and Ly-49H associate with mouse DAP12 and form activating receptors. J Immunol 161:7-10

Tomasello E, Olcese L, Vely F, Geourgeon C, Blery M, Moqrich A, Gautheret D, Djabali M, Mattei MG, Vivier E (1998) Gene structure, expression pattern, and biological activity of mouse killer cell activating receptor-associated protein (KARAP)/DAP-12. J Biol Chem 273:34115-34119

Wagtmann N, Biassoni R, Cantoni C, Verdiani S, Malnati MS, Vitale M, Bottino C, Moretta L, Moretta A, Long EO (1995) Molecular clones of the p58 NK cell receptor reveal immunoglobulin-related molecules with diversity in both the extra- and intracellular domains. Immunity 2:439-449

Wagtmann N, Rojo S, Eichler E, Mohrenweiser H, Long EO (1997) A new human gene complex encoding the killer cell inhibitory receptors and related monocyte/macrophage receptors. Curr Biol 7:615-618

Wilson MJ, Torkar M, Haude A, Milne S, Jones T, Sheer D, Beck S, Trowsdale J (2000) Plasticity in the organization and sequences of human KIR/ILT gene families. Proc Natl Acad Sci USA 97:4778-4783

Yamashita Y, Fukuta D, Tsuji A, Nagabukuro A, Matsuda Y, Nishikawa Y, Ohyama Y, Ohmori H, Ono M, Takai T (1998) Genomic structures and chromosomal location of p91, a novel murine regulatory receptor family. J Biochem (Tokyo) 123:358-368

14
The Fc receptor family structure based strategies for the development of anti-inflammatory drugs

P. Mark Hogarth , Maree S. Powell , Lisa J. Harris , Bruce Wines , and Gary Jamieson

The Austin Research Institute, Kronheimer Building, A&RMC Austin Campus, Studley Rd, Heidelberg, Victoria, 3084 Australia

Summary. Leukocyte Fc receptors play key roles in the initiation of antibody-mediated cellular responses to pathogens. However, they are also involved in the pathogenesis of destructive immune complex inflammatory diseases. The solving of the FcR crystallographic structures together with molecular modelling has provided an opportunity to examine closely the mode of interaction between receptor and ligand. Also such work has allowed the design of novel anti-inflammatory agents that act to block receptor:ligand interactions. These are potentially effective therapeutic agents in the treatment of immune complex mediated inflammatory diseases.

Keywords. Fc receptors, Immunoglobulin, Structure based design, Anti-inflammatory agents

Introduction

Receptors for immunoglobulins (Fc receptors) are found on many cell types. Most are members of the Ig superfamily of proteins with over 10 Ig-like Fc receptor proteins described (Hulett and Hogarth 1994a). Of particular interest in pathological immunity are the leukocyte receptors, FcγRI, FcγRII, FcγRIII, FcεRI, FcαRI, that are specific for IgG, IgE and IgA respectively. These receptor proteins exhibit extensive homology, with the extracellular domains all derived from the Ig superfamily domains. With the exception of FcγRI, which has three Ig-like domains, all have two such domains (Fig. 1). Despite their obvious homology at the level of amino acid sequence, they exhibit quite different specificity for ligand and very different affinities. Moreover the mechanisms by which these differences in specificity and affinity are generated are different for the different receptor classes. The solving of the three dimensional structure of the Fc receptors allows

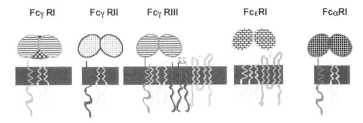

Fig. 1. Diagramatic representation of the leukocyte Fc receptors belonging to the Ig superfamily. These membrane proteins are shown along with the associated signalling subunits including the gamma subunit of FcεRI shared amongst many Fc recpetors as well as the beta subunit of FcεRI. All possible combinations of associated subunits are shown with FcγRIIIa, ie gama chain or zeta chain homodimers, as well as the gamma zeta heterodimer.

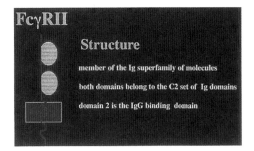

Fig. 2. Diagramatic representation of human FcγRIIa.

for the first time detailed analysis of interactions between Fc receptors and their immunoglobulin ligands. A knowledge of these will pave the way for the development of novel drugs to antagonise receptor function with the aim of ameliorating the affects of destructive inflammation caused by antibodies. Target diseases for the development of such anti-inflammatory drugs include:

(1) rheumatoid arthritis, where immune complexes play a major role in extra articular and joint disease,

(2) immune thrombocytopaenia, where platelet destruction is mediated by Fc receptor based mechanisms,

(3) allergy, where the classic type I hypersensitivity is a direct result of the cross linking of IgE bound to FcεRI.

X-ray crystallographic structures of Fc receptors

Our initial studies defined the structure and function of the most widespread of the activating human FcR - FcγRIIa (Fig. 2). In these studies we determined the structure of the extracellular domains of FcγRIIa (Maxwell et al. 1999). This receptor has two extracellular Ig-like domains joined at an angle of 52Å which is maintained by a network of hydrogen bonds between the domains. There are two major implications of this arrangement: first domain 1 is close to the membrane; second the Ig binding site, which has previously been defined by mutagenesis studies (Hulett et al. 1993, 1994a,b), occurs at the "top" of the second domain. Domain 1 is not involved in the interaction with IgG.

As the amino acid sequence of FcγRII is homologous to that of FcγRI, FcγRIII and FcεRI and homology modelling indicates that these Fc receptors have a similar overall structure (Figs. 3,4) (Rigby et al. 2000a,b). Indeed mutagenesis studies of FcεRI (Hulett et al. 1994b) and FcγRIII (Tamm and Schmidt 1997) indicate the location of the IgE and IgG binding sites in domain 2 in the same areas identified for FcγRIIa (Fig. 5). These homology models have been confirmed and extended by recent X-ray crystallographic analyses of other Fc receptors (Garman et al. 1998; Sondeman et al. 1999), including solution of the structure of complexes between FcεRI and FcγRIII with their ligands (Sonderman et al. 2000; Garman et al. 2000). These analyses have confirmed the contact residues, identified for the most part in mutagenesis studies (Hulett et al. 1993, 1994b, 1995; Tamm and Schmidt 1997). The regions that contribute to Fc receptor Ig interactions are largely composed of the areas comprising the BC loop, the C'E loop and the FG loop, with some contributions of the adjacent strands (Fig. 5).

Fig. 3. Ribbon diagram and Connelly surface of FcεRI as proposed in the homology model of Rigby et al. (2000a). Domains 1 and 2 are indicated. The IgE binding surface is coloured in aqua. The alpha carbon backbone shown in red.

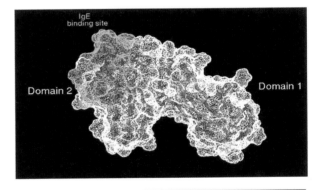

Fig. 4. Ribbon diagram and Connelly surface of human FcγRIIIb extracellular domains. Domain 2 shown on the left, domain 1 on the right, and the surface identified by limited mutagenesis as involved in IgG binding shown in yellow. The alpha carbon backbone is shown in green.

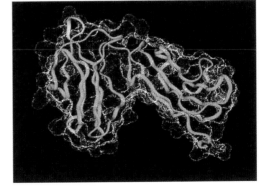

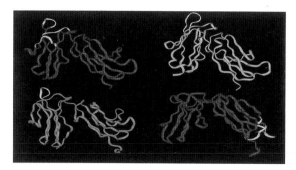

Fig. 5. Ribbon diagram of the alpha carbon backbones of human Fc Receptors. In each model domain 2 is shown on the left and domain 1 on the right. The ligand binding site is shown in yellow. The models shown clockwise from top left corner are: FcγRIIa (X-ray structure determined); FcεRI; FcαRI and FcγRIIIb.

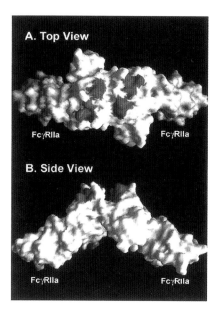

A. Top View

FcγRIIa   FcγRIIa

B. Side View

FcγRIIa   FcγRIIa

Fig. 6. Surface of FcγRIIa dimer viewed from the top, ie looking down the two-fold axis of symmetry, and from the side. Each dimer is annotated with the individual FcγRIIa monomers. The areas shown in red indicate the surface contributed by residues which have been mutated and shown to affect IgG binding. The groove and pocket that run across the centre of the dimer between the binding surface of each monomer is apparent in the middle of the dimer surface viewed on top and from the side.

Unique dimeric structure of FcγRIIa

Unlike the structures of FcεRI, FcγRIIb and FcγRIII, human FcγRIIa appears as a crystallographic dimer. Such a dimer presents an interesting complex with significant biological implications. In the dimer, the binding site of domain 2 of each monomer is brought together to form what is essentially a single contiguous surface (Fig. 6). Such a dimer has not been observed in other Fc receptors, and in the crystallographic complexes of FcγRIII and FcεRI no dimer was observed. Whilst the mutagenesis studies of FcγRIII and FcγRII identify similar areas involved in interaction with IgG, the interaction of IgG with a FcγRIIa dimer

Fig. 7. Model of the possible interaction of IgG2a with FcγRIIa dimer. The alpha carbon traces of the mouse IgG2a monoclonal antibody 231 is shown with its lower hinge and Cγ2 docked onto the alpha carbon trace of the FcγRIIa dimer. The contact areas would be expected to include those shown on the upper surface of the dimer, indicated in Fig. 6.

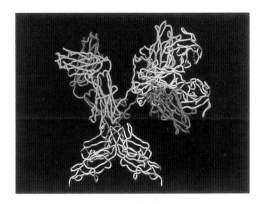

Fig. 8. Diagramatic representation of the FcγRIIa dimer viewed from the side, indicating the target areas for rational drug design in the groove created by the dimerisation of FcγRIIa, as indicated.

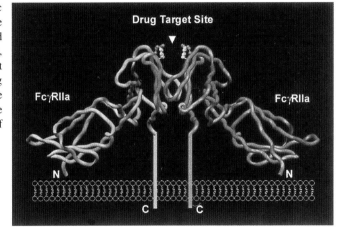

cannot occur in the same way as with FcγRIII monomer due to steric restrictions. FcγRIIa dimer formation would be prevented through a clash with other areas of the heavy chain not directly involved in binding. It would therefore appear that an interaction of IgG with dimeric FcγRIIa represents a novel mode of interaction with some similarities to other Fc receptors (Fig. 7).

Design of anti-inflammatories

The role of FcγRs as useful inflammatory targets was validated in unique studies using recombinant solube FcγRs to treat type III hypersensitivity reactions *in vivo* (Ierino et al. 1993). As a result of these studies we have targeted Fc receptors generally and FcγRIIa in particular for the development of anti-inflammatory agents.

The use of structure based drug design has successfully resulted in the development of a number of drugs available to treat disease. Most notable of these are the anti-influenza drug Relenza, a neuraminidase inhibitor, and the anti–AIDS drugs that target HIV protease. These drugs target enzyme active sites which are small, chemically and geometrically well defined, active sites. By contrast, the interactions between protein receptors and protein ligands usually involves large featureless surfaces, which makes the design of small chemical entities as potent antagonists of that interaction difficult.

The FcγRIIa dimer structure (Figs. 6,8) provides ideal features for the targeting and design of small chemical entities. As described above, the dimerisation of the receptor brings together the binding surfaces of each monomer to form a contiguous surface. As a result of the dimerisation a groove is formed that traverses the binding site. This well defined groove also contains a small pocket (about the two-fold axis of symmetry). The groove and pocket provide an ideal area to target the design of small chemical entities with the appropriate chemical properties to enable interaction with the side chains that protrude into and around the groove and the pocket. Although beyond the scope of this presentation, such small chemical entities have been designed and antagonise receptor induced cell activation (data not shown).

It should also be noted that because of the homology with other receptors, the information derived from the analysis of the 3-D structure of FcγRIIa would be applicable to other receptor systems. Whilst the extent of the conservation of dimerisation is a feature of Fc receptors is still to be determined, there are other areas of the receptor that are worthy of targeting, based on their structural features and conservation amongst the Fc receptor family. Thus we are targetting additional sites on FcγRIIa as well as other Fc receptors including FcεRI for the development of anti-inflammatories to treat a variety of autoimmune and pathological immune conditions.

The IgA receptor has unique structural characteristics

FcαRI (CD89) is an universal leukocyte Fc receptor having approximately 20% amino acid identity to Fcγ receptors, but in contrast approximately 30% identity to the killer inhibitory receptors (Wines et al. 1999). Most importantly FcαRI contains key structural features found in the killer inhibitory receptors but not in the Fc receptors, the most important of which is the linker between domains 1 and 2. The 5 amino acid linker of the killer inhibitory receptors is conserved in FcαRI, compared to the 2 amino acid linker of FcγRII, FcγRIII and FcεRI. The effect of this linker in FcαRI is to allow the the relationship of domains 1 and 2 to be very different to that in the other leukocyte Fc receptors. Domain 1 is effectively "rotated" 160 degrees in its relationship to domain 2, when compared to the same relationship in the other leukocyte Fc receptors (Wines et al. 1999, 2001) (Fig. 5).

Mutagenesis studies of FcαRI also revealed a different location of the ligand binding site for this receptor when compared to the other leukocyte Fc receptors. The ligand binding site of FcαRI is located at the tip of domain 1, not domain 2. However, the loops involved in domain 1 of FcαRI are the same as those involved in domain 2 of the other Fc receptors, implying some general conservation of ligand-recognition structure.

As the location of the binding site of FcαRI is different to that of the FcγR and FcεRI, it would be expected that the FcαR binding site of IgA would likewise be located elsewhere. Indeed whereas the receptor binding site of IgG and IgE is located in the lower hinge (or its equivalent Cε2), the FcαRI binding site is found in the interface between Cα2 and Cα3 (Carayannopoulos et al. 1996). This region would be the equivalent of the protein A binding site of IgG and is quite distinct from the lower hinge equivalent of IgG. Thus the mode of interaction of IgA with its receptor is different to that of IgG and raises the possibility that two FcαRI molecules can simultaneously contact one IgA molecule to each of the two IgA heavy chains (Wines et al. 2001). Thus the interaction between IgA and its receptor is clearly different to that of IgG and IgE and their receptors.

Conclusion

The development of new anti-inflammatories based on a knowledge of receptor structure and the mode of interaction between receptor and ligand will pave the way for new approaches for the development of anti-inflammatory drugs, and may provide convenient, cheap, cost-effective therapies for the treatment of many debilitating diseases.

References

Carayannopoulos L, Hexham JM, Capra JD (1996) Localization of the binding site for the monocyte immunoglobulin IgA-Fc receptor (CD89) to the domain boundary between Cα2 and Cα3 in human IgA1. J Exp Med 183:1579-1586

Garman SC, Kinet JP, Jardetzky TS (1998) Cyrstal structure of the human high affinity IgE receptor. Cell 95:951-961

Garman SC, Wuzburg BA, Tarchevskaya SS, Kinet JP, Jardetzky TS (2000) Structure of the Fc fragment of human IgE bound to its high-affinity receptor FcεRα. Nature 406:259-266

Hulett MD, Hogarth PM (1994a) The Fc receptors: their diversity of structure and function. Adv Immunol 57:1-127

Hulett MD, McKenzie IFC, Hogarth PM (1993) Chimeric Fc receptors identify immunoglobulin-binding regions in human FcγRII and FcεRI. Eur J Immunol 23:640-645

Hulett MD, Witort E, Brinkworth RI, McKenzie IFC, Hogarth PM (1994b) Identification of the IgG binding site of the low affinity receptor for IgG FcγRII: Enhancement and ablation of binding by site-directed mutagenesis. J Biol Chem 269:15287-15293

Ierino F, Powell MS, McKenize IFC, Hogarth PM (1993) Recombinant soluble human FcγRII: Production, characterisation and inhibition of the Arthus reaction. J Exp Med 178:1617-1628

Maxwell K, Powell MS, Hulett MD, Barton OA, McKenzie IFC, Garrett TPJ, Hogarth PM (1999) Crystal structure of the human leukocyte Fc receptor, FcγRIIa. Nature Struct Biol 6:437-442

Rigby LJ, Epa VC, Mackay G, Hulett MD, Sutton BJ, Gould HJ, Hogarth PM (2000a) Domain one of the high affinity IgE receptor, FcεRI, regulates binding to IgE through its interface with domain two. J Biol Chem 275:9664-9672

Rigby LJ, Trist H, Epa VC, Snider J, Hulett MD, Hogarth PM (2000b) Monoclonal antibodies and synthetic peptides define the active site of FcεRI and a potential receptor antagonist. Allergy 55:609-619

Sondermann P, Huber R, Jocob V (1999) Crystal structure of the soluble form of the human Fc receptor IIb. A new member of the Ig superfamily at 1.7Å resolution. EMBO J 18:1095-1103

Sondermann P, Huber R, Oosthuizen V, Jacob V (2000) The 3.2-Å crystal structure of the human IgG1 Fc fragment – FcγRIII complex. Nature 406:267-273

Tamm A, Schmidt RE (1997) IgG binding sites on human Fc receptors. Int Rev Immunol 16:57-85

Wines BD, Hulett MD, Jamieson GP, Trist HM, Spratt JM, Hogarth PM (1999) Identification of residues in the first domain of human Fcα receptors essential for interaction with IgA. J Immunol 162:2146-2153

Wines BD, Sardjono CT, Trist HM, Lay C-S, Hogarth PM (2001) Structural analysis of the interaction of FcαR with IgA and its implications for ligand binding by immunoreceptors of the LRC. J Immunol (in press)

Part C

Signaling Events and Physiological Roles

15
The mouse gp49 family

Howard R. Katz

Department of Medicine, Harvard Medical School and Division of Rheumatology, Immunology and Allergy, Brigham and Women's Hospital, Boston, MA 02115, USA

Summary. gp49A and gp49B1 are the products of highly homologous genes and are expressed in naïve animals on mast cells and macrophages. Expression is inducible on natural killer (NK) cells by culture in interleukin (IL)-2 in vitro or viral infection in vivo. gp49B1 contains two immunoreceptor tyrosine-based inhibitory motifs (ITIMs) in its cytoplasmic domain and inhibits mast cell and NK cell activation. The ITIMs are necessary to recruit the cytosolic, src homology 2 domain-containing tyrosine phosphatase SHP-1, and active enzyme is required for inhibition of mast cell activation by gp49B1. gp49A bears neither ITIMs, known activation motifs, nor sequences that promote association with other activating molecules. However, gp49A can function as an activating receptor by dimerizing and migrating to lipid rafts, which are enriched for signal transducing molecules. In addition to the gp49s, mast cells express activating and inhibitory receptors of the Paired Immunoglobulin (Ig)-like receptor (PIR) and IgG-Fc receptor (FcγR) families, suggesting that mast cell activation is regulated by a more intricate network of inhibitory and activating signals than previously appreciated.

Key words. gp49B1, gp49A, Immunoreceptor tyrosine-based inhibitory motif, Mast cells, NK cells

Expression of gp49 family members

The gp49 family was originally defined by the rat monoclonal antibody (mAb) B23.1, which detected an epitope on the surface of mouse macrophages and mast cells (LeBlanc et al. 1982; Katz et al. 1983). Subsequent studies revealed that, although mAb B23.1 does not bind to mouse NK cells from naïve animals, expansion of splenic NK cells by culture in IL-2 or stimulation in vivo by infection with cytomegalovirus elicits recognition by mAb B23.1 (Rojo et al. 1997; Wang et al. 1997; Wang et al. 2000). Immunochemical analyses of mast cells revealed the mAb B23.1 epitope is expressed on a 49-kDa glycoprotein (Katz et al. 1989). cDNA cloning, based on the amino-terminal sequence of the immunoreactive 49-kDa protein, revealed a family of highly homologous gp49 molecules, termed gp49A (Arm et al. 1991), gp49B1, and gp49B2 (Castells et al. 1994), each of which has two immunoglobulin (Ig)-like domains (Fig. 1). gp49A is encoded by the gp49A gene (McCormick et al. 1999), whereas gp49B1 and gp49B2 are the products of alternate mRNA splicing of the gp49B gene (Castells et al. 1994). The coding sequences of gp49A and gp49B1 predict transmembrane proteins, but gp49B2 may be a soluble form, because exon 6 that encodes the transmembrane domain is absent. As assessed with a gp49A-specific antibody, the molecule is expressed on mast cells in tandem with gp49B1, which is specifically recognized by mAb B23.1 (Katz et al. 1996; McCormick et al. 1999). Expression of both gp49A and gp49B1 is lower on mature serosal mast cells compared with immature bone marrow-derived mast cells (BMMC) (McCormick et al. 1999). mRNA encoding gp49A is elevated in NK cells from virus-infected mice, suggesting that the expression of gp49A and gp49B1 also change in parallel in NK cells. Expression of mRNA encoding gp49A and/or gp49B1 has also been detected in uterine endometrium at day 4 of pregnancy (Matsumoto et al. 1997).

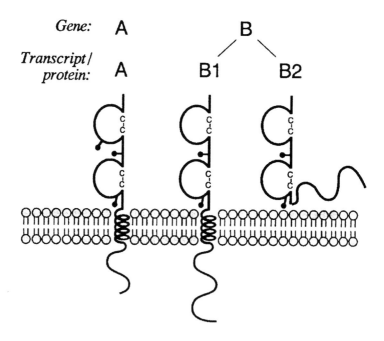

Fig. 1. The gp49 family.

Inhibition of cell activation by gp49B1

Of the two transmembrane gp49 molecules, only gp49B1 has two ITIMs in its cytoplasmic domain. Coligation of gp49B1 with the high affinity IgE-Fc receptor (FcεRI) on the surface of BMMC inhibits the release of secretory granule mediators and leukotriene C₄ (Katz et al. 1996), demonstrating a functional ITIM response and that gp49B1 is an endogenous inhibitor of FcεRI-initiated mast cell activation. Expression of a chimeric molecule bearing the cytoplasmic domain of gp49B1 in an NK cell line inhibits NK cell functions, including cytokine release (Rojo et al. 1997; Wang et al. 2000).

The contributions of the gp49B1 ITIM tyrosines to inhibition of mast cell activation have been assessed by expressing wild-type gp49B1 and gp49B1 with tyrosine-to-phenylalanine mutations in either or both ITIMs in the rat basophilic leukemia-2H3 (RBL) cell line (Lu-Kuo et al. 1999). Mutation of either ITIM tyrosine partially attenuates the ability of gp49B1 to inhibit exocytosis, whereas alteration of both tyrosines abolishes suppression, demonstrating the critical contribution of the ITIM tyrosines to gp49B1-mediated inhibition of mast cell activation. Sodium pervanadate elicits the phosphorylation of native gp49B1 in BMMC, as well as association of cytosolic SHP-1 and SHP-2 with gp49B1. However, there is no association of the SH2 domain-containing inositol polyphosphate 5'-phosphatase SHIP, which associates with FcγRIIB and mediates cellular inhibition in mast cells and other cells of the immune system (Ono et al. 1996). In contrast with the results obtained with pervanadate, only SHP-1 associates with gp49B1during attenuation of FcεRI-induced mast cell activation (Lu-Kuo et al. 1999). Similar interactions occur when gp49B1 is expressed in NK cells (Wang et al. 1999). Importantly, inhibition of FcεRI-mediated activation by gp49B1 is reduced in BMMC generated from mice with defective SHP-1, demonstrating that the inhibitory effect of gp49B1 is dependent on the enzymatic activity of SHP-1 (Lu-Kuo et al. 1999). gp49B1 inhibits both the FcεRI-induced release of calcium from intracellular stores and the subsequent influx of extracellular calcium in BMMC (Lu-Kuo et al. 1999), the latter of which is required for mast cell exocytosis and leukotriene generation. Hence, the release of two classes of proinflammatory mediators from mast cells is inhibited by gp49B1 in an ITIM-, SHP-1-, and calcium-dependent manner.

Cell activation by gp49A

The gp49A and gp49B1 cDNAs are ~97% identical in their nucleotide sequences, with 88% amino acid identity (Castells et al. 1994). However, exon 8 of the gp49A gene does not have two adjacent cytosine residues that are present in the gp49B gene. This difference results in a translation reading frame shift in gp49A that causes an earlier stop codon relative to gp49B1 and absence of the two ITIMs (Arm et al. 1991; Castells et al. 1994; McCormick et al. 1999). The highly homologous forms of gp49 molecules, with and without ITIMs, are similar to those in related families of molecules, namely, the mouse Paired Ig-like Receptors (PIRs), human Killer cell Ig-like Receptors (KIRs), and human Leukocyte Ig-like receptors (LIRs). The non-ITIM-bearing molecules in all of these families do not have signaling motifs in their short cytoplasmic domains. However, all but gp49A have a positively charged amino acid in the transmembrane domain that facilitates association with DAP12 (Lanier et al. 1998) or the FcR γ-chain (Nakajima et al. 1999). These subunits transduce activation signals through immunoreceptor tyrosine-based activation motifs. Nonetheless, chimeric receptors consisting of the extracellular domain of FcγRIIB fused to the membrane-proximal, transmembrane, and cytoplasmic domains of gp49A stimulate calcium flux as well as prostaglandin D2 and cytokine generation when cross-linked on the surface of RBL transfectants (Lee et al. 2000). These responses are associated with homodimerization of the chimeric gp49A molecules through an extracellular cysteine proximal to the plasma membrane. Moreover, the molecules migrate to lipid rafts, which are enriched for signaling molecules. These findings suggest that gp49A promotes cell activation through a novel signaling mechanism.

Genomic considerations

The gp49A (McCormick et al. 1999) and gp49B (Castells et al. 1994) genes have a striking, 94% nucleotide sequence identity. Indeed, the introns of the genes have as much nucleotide identity as the exons. The two genes are nearly contiguous, being within ~4.5 kb of each other on mouse chromosome 10 (Kuroiwa et al. 1998). These characteristics suggest that the two genes arose by duplication with relatively little subsequent mutation. The chromosomal localization of the gp49 genes is another unique feature of this family. Mouse chromosome 10 is not syntenic with human chromosome 19, where the KIR and LIR genes are located. In contrast, the genes encoding the PIR family are located on mouse chromosome 7 (Kubagawa et al. 1997; Yamashita et al. 1998), which is syntenic with human chromosome 19. The distinct chromosomal localization of the gp49 genes, and the absence of ITIMs and a charged transmembrane amino acid in gp49A, suggests that the gp49 family is the most structurally and functionally diverged family from among the PIRs, KIRs, and LIRs.

References

Arm JP, Gurish MF, Reynolds DS, Scott HC, Gartner CS, Austen KF, Katz HR (1991) Molecular cloning of gp49, a cell surface antigen that is preferentially expressed by mouse mast cell progenitors and is a new member of the immunoglobulin superfamily. J Biol Chem 266:15966-15973

Castells MC, Wu X, Arm JP, Austen KF, Katz HR (1994) Cloning of the gp49B gene of the immunoglobulin superfamily and demonstration that one of its two products is an early-expressed mast cell surface protein originally described as gp49. J Biol Chem 269:8393-8401

Katz HR, LeBlanc PA, Russell SW (1983) Two classes of mouse mast cells delineated by monoclonal antibodies. Proc Natl Acad Sci USA 80:5916-5918

Katz HR, Benson AC, Austen KF (1989) Activation- and phorbol ester-stimulated phosphorylation of a plasma membrane glycoprotein antigen expressed on mouse IL-3-dependent mast cells and serosal mast cells. J Immunol 142:919-926

Katz HR, Vivier E, Castells MC, McCormick MJ, Chambers JM, Austen KF (1996) Mouse mast cell gp49B1 contains two immunoreceptor tyrosine-based inhibition motifs and suppresses mast cell activation when coligated with the high-affinity Fc receptor for IgE. Proc Natl Acad Sci USA 93:10809-10814

Kubagawa H, Burrows PD, Cooper MD (1997) A novel pair of immunoglobulin-like receptors expressed by B cells and myeloid cells. Proc Natl Acad Sci USA 94:5261-5266

Kuroiwa A, Yamashita Y, Inui M, Yuasa T, Ono M, Nagabukuro A, Matsuda Y, Takai T (1998) Association of tyrosine phosphatases SHP-1 and SHP-2, inositol 5-phosphatase SHIP with gp49B1, and chromosomal assignment of the gene. J Biol Chem 273:1070-1074

Lanier LL, Corliss BC, Wu J, Leong C, Phillips JH (1998) Immunoreceptor DAP12 bearing a tyrosine-based activation motif is involved in activating NK cells. Nature 391:703-707

LeBlanc PA, Russell SW, Chang S-MT (1982) Mouse mononuclear phagocyte heterogeneity detected by monoclonal antibodies. J Reticuloendothelial Soc 32:219-231

Lee KH, Ono M, Inui M, Yuasa T, Takai T (2000) Stimulatory function of gp49A, a murine Ig-like receptor, in rat basophilic leukemia cells. J Immunol 165:4970-4977

Lu-Kuo JM, Joyal DM, Austen KF, Katz HR (1999) gp49B1 inhibits IgE-initiated mast cell activation through both immunoreceptor tyrosine-based inhibitory motifs, recruitment of the src homology 2 domain-containing phosphatase-1, and suppression of early and late calcium mobilization. J Biol Chem 274:5791-5796

Matsumoto Y, Handa S, Taki T (1997) gp49B1, an inhibitory signaling receptor gene of hematopoietic cells, is induced by leukemia inhibitory factor in the uterine endometrium just before implantation. Dev Growth Differ 39:591-597

McCormick MJ, Castells MC, Austen KF, Katz HR (1999) The gp49A gene has extensive sequence conservation with the gp49B gene and provides gp49A protein, a unique member of a large family of activating and inhibitory receptors of the immunoglobulin superfamily. Immunogenetics 50:286-294

Nakajima H, Samaridis J, Angman L, Colonna M (1999) Human myeloid cells express an activating ILT receptor (ILT1) that associates with Fc receptor γ-chain. J Immunol 162:5-8

Ono M, Bolland S, Tempst P, Ravetch JV (1996) Role of the inositol phosphatase SHIP in negative regulation of the immune system by the receptor FcγRIIb. Nature 383:263-266

Rojo S, Burshtyn DN, Long EO, Wagtmann N (1997) Type I transmembrane receptor with inhibitory function in mouse mast cells and NK cells. J Immunol 158:9-12

Wang LL, Mehta IK, LeBlanc PA, Yokoyama WM (1997) Mouse natural killer cells express gp49B1, a structural homology of human killer inhibitory receptors. J Immunol 158:13-17

Wang LL, Blasioli J, Plas DR, Thomas ML, Yokoyama WM (1999) Specificity of the SH2 domains of SHP-1 in the interaction with the immunoreceptor tyrosine-based inhibitory motif-bearing receptor gp49B. J Immunol 162:1318-1323

Wang LL, Chu DT, Dokun AO, Yokoyama WM (2000) Inducible expression of the gp49B inhibitory receptor on NK cells. J Immunol 164:5215-5220

Yamashita Y, Fukuta D, Tsuji A, Nagabukuro A, Matsuda Y, Nishikawa Y, Ohyama Y, Ohmori H, Ono M, Takai T (1998) Genomic structures and chromosomal location of p91, a novel murine regulatory receptor family. J Biochem 123:358-368

16
Regulation of B-cell antigen receptor signaling by CD72

Takeshi Tsubata, Chisato Wakabayashi, and Takahiro Adachi

Department of Immunology, Medical Research Institute, Tokyo Medical and Dental University, 1-5-45 Yushima, Bunkyo-ku, Tokyo 113-8510, Japan

Summary. CD72 is a B cell membrane molecule containing an immunoreceptor tyrosine-based inhibition motif (ITIM) in the cytoplasmic region. CD72 constitutively associates with BCR and is phosphorylated upon BCR ligation, resulting in recruitment of the SHP-1 phosphatase. CD72-deficient B cells show enhanced ERK activation and calcium mobilization upon BCR ligation, indicating that CD72 negatively regulates BCR signaling probably by activating SHP-1, thereby setting a signaling threshold for B cell activation. Treatment with anti-CD72 antibody abrogates BCR ligation-induced phosphorylation of CD72, suggesting that anti-CD72 antibody activates B cells by inhibiting negative regulatory function of CD72. CD72 may thus regulate B cell activation as a negative regulator of BCR signaling.

Key words. CD72, SHP-1, BCR, ITIM

Introduction

Several B cell membrane molecules containing immunoreceptor tyrosine-based inhibition motifs (ITIMs) have been shown to negatively regulate BCR signaling (Tsubata 1999). Among these molecules, the low affinity receptor for the Fc portion of IgG (FcγRII) and CD22 are most extensively characterized. CD22 constitutively associates with BCR, and negatively regulates BCR signaling whenever B cells interact with antigen, thereby setting a signaling threshold for B cell activation (Cyster and Goodnow 1997; O'Rourke et al. 1997; Tedder et al. 1997). In contrast, FcγRII associates with BCR and negatively regulates BCR signaling only when B cells interact with antigen-IgG immune complexes, resulting in negative regulation of IgG synthesis (Coggeshall 1998). These ITIM-containing molecules activate SH2-containing phosphatases such as SH2-

containing inosital polyphosphate 5-phosphatase (SHIP) and SH2-containing protein tyrosine phosphatase 1 (SHP-1), which in turn negatively regulate BCR signaling probably by dephosphorylating signaling molecules.

CD72 is a 45 kD type II membrane molecule expressed on B cells as a homodimer (Nakayama et al. 1989). Treatment with anti-CD72 antibody has been shown to activate B cells (Subbarao and Mosier 1982; Yakura et al. 1982), suggesting that CD72 regulates B cell activation. CD72 contains a C-type lectin-like domain in the extracellular part and an ITIM in the cytoplasmic region. Here we describe recent advance in understanding the regulatory role of CD72 in B cell activation.

Association of CD72 with SH2-containing phosphatases

In addition to an ITIM, CD72 contains an ITIM-like sequence in the cytoplasmic region. We generated a GST fusion protein containing either the ITIM or the ITIM-like sequence of CD72 and demonstrated that, upon tyrosine phosphorylation, the ITIM but not the ITIM-like sequence binds to SHP-1 *in vitro* (Adachi et al. 1998). Hozumi and his colleagues obtained essentially the same result by using synthetic peptides (Wu et al. 1998). They further demonstrated that the phosphorylated ITIM-like sequence binds to the adaptor molecule Grb2. These findings indicate that the ITIM in CD72 is function and is capable of recruiting the SH2-containing phosphatase SHP-1, whereas the ITIM-like sequence in CD72 recruits Grb2.

The association of CD72 with SHP-1 appears to take place in B cells. Both our group and Hozumi's group showed that CD72 is phosphorylated and is co-precipitated with SHP-1 in the BCR-ligated B cell line WEHI-231 (Adachi et al. 1998; Wu et al. 1998). In contrast, SHIP was not co-precipitated with CD72, although expressed in WEHI-231 cells. These results suggest that, upon tyrosine phosphorylation, CD72 recruits SHP-1 but not SHIP in B cells. Moreover, BCR ligation-induced CD72 phosphorylation suggests that CD72 is associated with BCR and is phosphorylated by BCR-associated kinases such as Lyn when they are activated by BCR ligation. Physical association between CD72 and BCR is supported by our observation that CD72 is specifically co-precipitated with BCR (Adachi et al. unpublished). Taken together, CD72 associates with BCR constitutively, and, upon BCR ligation, recruits SHP-1 probably at the phosphorylated ITIM. This suggests that CD72 negatively regulates BCR ligation-induced B cell activation as SHP-1 negatively regulates cell activation.

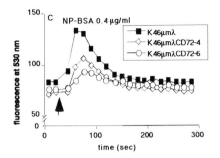

Fig. 1. CD72 negatively regulates BCR ligation-induced calcium flux in the B cell line K46μmλ cells expressing BCR reactive to the hapten NP. K46μmλ and its CD72 transfectants were stimulated with the antigen NP-coupled bovine serum albumin (BSA), and intracellular free Ca²⁺ concentration was monitored by using Fluo-3. Taken from Adachi et al. (2000) with permission.

Negative regulation of BCR signaling by CD72

To ask whether CD72 negatively regulates BCR signaling, both our group and Parnes' group compared BCR signaling between B cells expressing CD72 and those that lack CD72 (Adachi et al. 2000; Pan et al. 1999). The B cell line K46μmλ expresses surface IgM reacting to the hapten nitrophenol (NP) but does not express CD72. We established K46μmλ CD72 transfectants, and examined ERK activation and calcium response induced by treatment with the antigen NP because ERK and calcium signaling are essential for B cell activation induced by BCR ligation (Adachi et al. 2000). Antigen-induced ERK activation and calcium response were significantly reduced in the CD72 transfectants, indicating that CD72 down-modulates BCR signaling for B cell activation. Parnes and her colleagues established mouse lines deficient in CD72 (Pan et al. 1999). By comparing CD72-deficient B cells with normal B cells, they demonstrated the results similar to those obtained in K46 cells. BCR ligation induces enhanced ERK activation and calcium mobilization in CD72-deficient B cells compared to normal B cells.

We demonstrated that CD72 requires the ITIM for negatively regulating BCR signaling in K46 cells. Thus, recruitment of SHP-1 appears to be essential for CD72 to negatively regulate BCR signaling. SHP-1 activated by CD72 may regulate BCR signaling by dephosphorylating signaling molecules activated by BCR ligation. Upon BCR ligation, cytoplasmic protein tyrosine kinases such as Syk are recruited to BCR and activated (Kurosaki 1999; Reth and Wienands 1997). Syk then phosphorylates the adaptor molecule SLP-65/BLNK, thereby activating phospholipase Cγ essential for calcium response mediated by BCR (Adachi et al. unpublished). Mizuno et al. demonstrated that SHP-1 associates with SLP-65/BLINK in BCR-ligated WEHI-231 cells (Mizuno et al. 2000). They further demonstrate that expression of a dominant negative form of SHP-1 enhances phosphorylation of SLP-65/BLNK but not the upstream kinase Syk. Based on these observations, they proposed that SLP-65/BLINK is the substrate of SHP-1 for negative regulation of BCR signaling. In contrast, we demonstrated that SHP-1 activated by CD72 negatively regulates phosphorylation of Syk as well as SLP-65/BLNK by using a myeloma line expressing both BCR and CD72. Our result is consistent with the previous observation that expression of a dominant negative mutant of SHP-1 enhances Syk phosphorylation in B cells (Dustin et al. 1999). Thus, SHP-1 activated by CD72 appears to dephosphorylate BCR-proximal signaling molecules such as Syk and/or SLP-65/BLNK. However, further studies are required to elucidate the direct substrate of SHP-1 for negative regulation of BCR signaling.

Regulation of B cell activation and apoptosis by CD72

Treatment with anti-CD72 antibodies has been shown to activate B cells (Subbarao and Mosier 1982; Yakura et al. 1982). One possible explanation is that this treatment induces dissociation of CD72 from BCR. Indeed, we demonstrated that BCR ligation-induced CD72 phosphorylation is inhibited almost completely in the presence of anti-CD72 antibodies (Adachi et al. unpublished). However, Bondada and his colleagues proposed that CD72 transmit a positive signal independently of BCR, based on the observation that strong Syk phosphorylation is induced by BCR ligation but not treatment with anti-CD72 antibody (Venkataraman et al. 1998). However, no direct evidence so far supports the positive signaling function of CD72.

We previously showed that treatment with anti-CD72 antibody blocks B cell apoptosis induced by BCR ligation (Nomura et al. 1996). This effect of anti-CD72 antibody cannot be explained by the model where CD72 negatively regulates BCR signaling for apoptosis as well as activation, because this model suggests that enhanced BCR signaling by anti-CD72 antibody should enhance BCR-mediated apoptosis. CD72 may regulate B cell apoptosis independently of its regulation on BCR signaling although the detailed mechanism is not yet known.

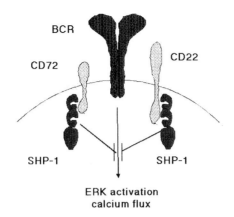

Fig. 2. Model of co-receptor-mediated negative regulation of BCR signaling. Both CD22 and CD72 associate with BCR constitutively. Whenever B cell interact with antigens, these inhibitory co-receptors negatively regulate BCR signaling by recruiting and activating the SHP-1 phosphatase.

Conclusion

Several lines of evidence now strongly suggest that CD72 constitutively associates with BCR on B cell surface, and negatively regulates BCR signaling whenever BCR is ligated by antigens. This negative regulation of BCR signaling appears to be mediated by SHP-1. These features are shared by another inhibitory co-receptor CD22, suggesting that CD72 plays a role redundant with that of CD22 in regulating BCR signaling. This notion is consistent with the observation that CD22-deficient mice show a marginal phenotype compared to marked B cell hyperactivity in SHP-1-deficient mice.

SHP-1 might be involved in CD72-mediated regulation of B cell apoptosis. Since SHP-1-deficiency causes development of autoimmune disease, it would be possible that CD72 regulates self-reactive B cells by controlling B cell apoptosis. Further studies are required for fully elucidating the roles of CD72 in B cell physiology.

References

Adachi T, Flaswinkel H, Yakura H, Reth M, Tsubata T (1998) The B cell surface protein CD72 recruits the tyrosine phosphatase SHP-1 upon tyrosine phosphorylation. J Immunol 160: 4662-4665

Adachi T, Wakabayashi C, Nakayama T, Yakura H, Tsubata T (2000) CD72 negatively regulates signaling through the antigen receptor of B cells. J Immunol 164: 1223-1229

Coggeshall KM (1998) Inhibitory signaling by B cell FcγRIIb. Curr Opin Immunol 10: 306-312

Cyster JG, Goodnow CC (1997) Tuning antigen receptor signaling by CD22: integrating cues from antigens and the microenvironment. Immunity 6: 509-517

Dustin LB, Plas DR, Wong J, Hu YT, Soto C, Chan AC, Thomas ML (1999) Expression of dominant-negative src-homology domain 2-containing protein tyrosine phosphatase-1 results in increased Syk tyrosine kinase activity and B cell activation. J Immunol 162: 2717-2724

Kurosaki T (1999) Genetic analysis of B cell antigen receptor signaling. Annu Rev Immunol 17: 555-592

Mizuno K, Tagawa Y, Mitomo K, Arimura Y, Hatano N, Katagiri T, Ogimoto M, Yakura H (2000) Src homology region 2 (SH2) domain-containing phosphatase-1 dephosphorylates B cell linker protein/SH2 domain leukocyte protein of 65 kDa and selectively regulates c-Jun NH_2-terminal kinase activation in B cells. J Immunol 165: 1344-1351

Nakayama E, von Hoegen I, Parnes JR (1989) Sequence of the Lyb-2 B-cell differentiation antigen defines a gene superfamily of receptors with inverted membrane orientation. Proc Natl Acad Sci USA 86: 1352-1356

Nomura T, Han H, Howard MC, Yagita H, Yakura H, Honjo T, Tsubata T (1996) Antigen receptor-mediated B cell death is blocked by signaling via CD72 or treatment with dextran sulfate and is defective in autoimmunity-prone mice. Int Immunol 8: 867-875

O'Rourke L, Tooze R, Fearon DT (1997) Co-receptors of B lymphocytes. Curr Opin Immunol 9: 324-329

Pan C, Baumgarth N, Parnes JR (1999) CD72-deficient mice reveal nonredundant roles of CD72 in B cell development and activation. Immunity 11: 495-506

Reth M, Wienands J (1997) Initiation and processing of signals from the B cell antigen receptor. Annu Rev Immunol 15: 453-479

Subbarao B, Mosier DE (1982) Lyb antigens and their role in B lymphocyte activation. Immunol Rev 69: 81-97

Tedder T. F. Tuscano J. Sato S. Kehrl J. H. (1997). CD22, a B lymphocyte-specific adhesion molecule that regulates antigen receptor signaling. Annu Rev Immunol 15: 481-504.

Tsubata T (1999). Co-receptors on B lymphocytes. Curr Opin Immunol 11: 249-255

Venkataraman C, Muthusamy N, Muthukkumar S, Bondada S (1998) Activation of lyn, blk, and btk but not syk in CD72-stimulated B lymphocytes. J Immunol 160: 3322-3329

Wu Y, Nadler MJ, Brennan LA, Gish GD, Timms JF, Fusaki N, Jongstra BJ, Tada N, Pawson T, Wither J, Neel BG, Hozumi N (1998) The B-cell transmembrane protein CD72 binds to and is an in vivo substrate of the protein tyrosine phosphatase SHP-1. Curr Biol 8: 1009-1017

Yakura H, Shen FW, Bourcet E, Boyse EA (1982) Evidence that Lyb-2 is critical to specific activation of B cells before they become responsive to T cell and other signals. J Exp Med 155: 1309-1316

17
A role for the SH2-containing inositol phosphatase in the biology of natural killer cells and stem cells

Tomar Ghansah, John M. Ninos, and William G. Kerr[1]

Immunology Program, H. Lee Moffitt Cancer Center and Research Institute and the Institute for Biomolecular Science, University of South Florida, 12902 Magnolia Drive, Tampa, Florida 33612, USA

Summary. SHIP is an inositol 5' phosphatase that can hydrolyze $PI(3,4,5)P_3$, a product of PI3 Kinase. SHIP was initially identified as a 145kD tyrosine phosphorylated protein known to associate with the adapter protein Shc following cytokine stimulation. In addition, tyrosine phosphorylated SHIP can bind to ITIM motifs of receptors expressed by immune cells and thus modulate their effector functions. Ly49 receptors are immune receptors found on murine Natural Killer (NK) cells that control their effector function. NK cells play an important role in the innate immune system and acute bone marrow graft rejection. We have found that in the absence of SHIP expression, murine NK cells exhibit a dysregulated Ly49 repertoire with NK cells expressing multiple Ly49 receptors specific for self MHC ligands. This defect prevents adult NK cells from lysing tumor cells *in vitro*. We have also identified a novel stem cell-specific isoform of SHIP, s-SHIP, whose expression is restricted to embryonic and hematopoietic stem cells (HSC). Therefore, in addition to regulating myelopoeisis *in vivo* and effector functions of B cells and mast cells, SHIP also plays a significant role in the biology of NK cells and stem cells.

Key words. SHIP, Inositol phospholipid signaling, Ly49, Stem cells, Natural Killer cells

---

[1] Correspondence to: William G. Kerr, Ph.D., Room 4072E, H. Lee Moffitt Cancer Center and Research Institute, 12902 Magnolia Drive, Tampa, Florida 33612. Email: kerrw@moffitt.usf.edu

129

Introduction

Inositol phospholipid signaling plays a key role in controlling the growth and survival of cells. PI3 Kinase (PI3K) and various inositol phosphatases are important mediators of these pathways. A key inositol phosphatase in the hematopoietic system is the SH2-containing Inositol Phosphatase (SHIP). SHIP can hydrolyze key inositol phospholipids (such as PIP_3) that promote the survival and proliferation of hematopoietic cells. It has been speculated that SHIP is involved in numerous signal transduction pathways in hematopoietic cells. SHIP was initially identified as a 145-kD intracellular signal transduction molecule that is tyrosine phosphorylated and recruited to the plasma membrane in response to a wide variety of stimuli in hematopoietic cells (Damen et al. 1996; Kananaugh et al. 1996; Lioubin et al. 1996; Ono et al. 1996).

SHIP contains an SH2 domain located at its amino terminus with a 5' inositol phosphatase domain in the center and a proline-rich region at the carboxy terminus of the protein. PTB binding sites as well as PI3K binding sites can also be found at the COOH terminus (Rohrschneider and Lucas 1999).

Following ligand binding to either cytokine or growth factor receptors, PI3-kinase is activated which then phosphorylates the 3' position of $PI(4,5)P_2$ to yield $PI(3,4,5)P_3$ (Franke et al. 1997). $PI(3,4,5)P_3$ enables membrane recruitment of various kinases that contain pleckstrin homology (PH) domains such as Protein Kinase B (AKT/PKB) and Bruton's Tyrosine Kinase (Btk). SHIP is also tyrosine phosphorylated upon growth factor receptor engagement and translocated to the plasma membrane near its lipid substrate $PI(3,4,5)P_3$ (Liu et al. 1994). SHIP can then selectively remove the 5' phosphate from $PI(3,4,5)P_3$ or soluble $Ins(1,3,4,5)P_4$ to produce the phospholipid products $PI(3,4)P_2$ and Ins $(1,3,4,)P_3$ (Damen et al. 1996; Lioubin et al. 1996). This activity of SHIP can prevent membrane recruitment of Akt/PKB and thus prevent the survival of the cell. Alternatively, Btk recruitment and activation by $PI(3,4,5)P_3$ could lead to induction of a JNK1-mediated stress responses that results in apoptosis. By hydrolysis of $PI(3,4,5)P_3$, SHIP would prevent this stress-response and thus promote cell survival (Ono et al. 1997). The product of PIP_3 hydrolysis, $PI(3,4)P_2$, may also enable membrane recruitment of other PH domain containing signaling molecules such as phosphoinositide-dependent kinase 1 (PDK1), Grp-1, VaV and PLC-γ (Rameh and Cantley 1999; Huber et al. 1999). Thus, SHIP may act to either promote the survival or apoptosis of the cell in response to a wide variety of extracellular stimuli. Whether SHIP mediates survival or apoptosis likely depends upon the signal, the cell type or both.

Several SHIP isoforms have been identified with apparent molecular weights of 145, 135, 125 and 110 kD (Damen et al. 1998). Expression of these isoforms may vary with either cell type or the differentiation stage of the cell. Another SH2-containing inositol phosphatase, known as SHIP2, has been identified. This 155kD protein is expressed not only in hematopoietic cells but also in skeletal muscle, heart, placenta and pancreas (Pessess et al. 1997; Bruhns et al. 1999).

Interestingly, PIP$_3$ is the only known substrate for SHIP2 (Wisniewski et al., 1999). Most SHIP isoforms are tyrosine phosphorylated in response to cytokine or growth factor stimuli (Krystal, 2000; Rohrschneider and Lucas, 1999) although a stem cell specific isoform has recently been identified that does not appear to be phosphorylated in response to LIF or SCF (Tu et al. 1999; Tu et al., submitted). The tyrosine phosphorylated 145 kD and 135 kD isoforms of SHIP bind specifically to the adaptor protein called Shc. The 110 SHIP isoform appears only to associate with the cytoskeleton (Damen et al. 1998). Tyrosine phosphorylated SHIP can form a bidentate complex with Shc through its SH2 domain or with Grb2 via its proline-rich region. Whether Shc or Grb2 partners with SHIP may be dependent on the cell type, the different kinases present or the affinity of SHIP for its receptor (Liu et al. 1994). The 110, 135 and 145kD isoforms have only been shown to bind to Grb2 *in vitro*. However, a stem cell specific SHIP isoform, s-SHIP, has been identified that binds to Grb2 *in vivo* (Tu et al., submitted).

SHIP, via its SH2 domain, has affinity for certain receptors (such as, FcγRII, etc.) that contain Immunoreceptor Tyrosine-based Inhibition Motifs (ITIM) (Vely et al., 1997; Tridandapani et al. 1997). SHIP binds to the cytoplasmic tails of ITIM-bearing receptors expressed on myeloid, B, T and NK cells. Receptors with tyrosine phosphorylated ITIM motifs recruit to the plasma membrane molecules that have either tyrosine or inositol phosphatase activity such as SHP-1, SHP-2 or SHIP (Scharenberg and Kinet 1996). During cell activation the Immunoreceptor Tyrosine–based Activation Motifs (ITAM) of activating receptors are phosphorylated leading to the recruitment of non-receptor Src-like kinases (e.g., Syk and Zap-70 tyrosine kinase) (Leibson, 1997; Vivier and Daeron 1997).

SHIP binds to certain receptors (e.g., FcγRIIB1 and gp49B1) (Vely et al. 1997; Kuroiwa et al. 1998) at their tyrosine phosphorylated ITIMs through its SH2 domain (Osborne et al. 1996). The SH2 domain of SHIP can also bind to ITAM motifs of the FcγRIIa and FcγR γ chain of monocytes (Maresco et al. 1999), the FcεRI γ chain on mast cells and the CD3 complex of the T cell receptor (TcR) (Osborne et al. 1996). In addition, the SH2 domain of SHIP enables a bidentate interaction between SHIP and Shc that is proposed to prevent membrane recruitment Grb2-mSOS complexes and subsequent activation of the Ras/MAPK pathway (Tridandapani et al. 1997).

Mast cells

Mast cells express the FcεR1 receptor which binds immune complexes containing IgE. These immune complexes mediate the allergic hypersensitivity reaction. The FcεR1 receptor is composed of an α, β and two γ-subunits (Damen et al. 1996). The β and γ subunits are located at the cytoplasmic tail and contain ITAM that are essential for IgE-mediated mast cell activation (Damen et al. 1996). SHIP's role in mast cells was investigated by studying bone marrow-derived mast cell lines (BMMCs) derived from SHIP$^{-/-}$ knockout mice. SHIP$^{-/-}$ mast cells are more sensitive to IgE-mediated degranulation than SHIP$^{+/+}$ or SHIP$^{+/-}$ mast cells. Interestingly, IgE stimulates massive degranulationof SHIP$^{-/-}$, but not SHIP$^{+/+}$ mast cells. In addition, when SHIP$^{-/-}$ mast cells were stimulated with IgE in the presence of EGTA, there was a decrease in the degree of degranulation of SHIP$^{-/-}$ mast cells (Ono et al. 1997). Therefore, SHIP appears to limit calcium influx in mast cells and thereby prevents inappropriate degranulation of mast cells. *In vitro* studies showed that SHIP binds to the second ITIM motif of the gp49B1 receptor in mast cells (Katz et al. 1996).

Crosslinking of the FcεRI receptor together with the gp49B1 receptor inhibited granule release from bone barrow-derived mast cells (Katz et al. 1996). Gp49B1 belongs to the immunoglobulin receptor superfamily. There are two isoforms of gp49B1: A and B. Gp49A lacks ITIM motifs whereas gp49B contains several ITIM motifs. Gp49B1 can inhibit cellular activation pathways in mast cells (Castells et al. 1994). Initially, gp49B1 receptors were thought to be exclusive to mast cells, but are also found on NK cells as well (Wang et al. 1997).

B lymphocytes

B cell receptor (BCR) activation alone induces the activation of tyrosine kinases, increases the cytoplasmic concentration of Ins(1,4,5)P$_3$ and Ca^{2+} and activation of the MAPK pathway (DeFranco 1997). These signaling events trigger the activation, proliferation and differentiation of B lymphocytes. However, when the BCR and FcγRIIB are crosslinked, extracellular Ca^{2+} influx is inhibited resulting in a decrease in cell proliferation and an inhibition of B cell differentiation (Phillips and Parker 1984). SHIP binds to the ITIM of the FcγRIIB and is tyrosine phosphorylated in response to BCR-FcγRIIB co-ligation. Therefore, SHIP is thought to be a negative regulator of B cells in response to immune complexes. SHIP may mediate this role by preventing activation of the Ras/MAPK pathway. The recruitment of SHIP to FcγRIIB ITIM where it is tyrosine phosphorylated and associates with Shc may prevent activation of the Ras/MAPK pathway (Tridandapani et al. 1997). SHIP could also mediate negative signaling by preventing Ca^{2+} influx via hydrolysis of soluble Ins(1,3,4,5)P$_4$ to Ins(1,3,4)P$_3$. Ins(1,3,4)P$_3$ promotes the closing of Ca^{2+} channels on the membrane preventing the influx of extracellular Ca^{2+} (Ono et al. 1997).

Activation of FcγRIIB alone on B cells induces the activation of Btk that can activate *JNK1*-induced stress responses leading to apoptosis of the cell (Bollard and Ravetch 1998). However, crosslinking the BCR and FcγRIIB receptors results in the activation of Syk and Lyn, the phosphorylation of the FcγRIIB ITIM and recruitment of SHIP. Membrane recruitment of SHIP promotes B cell survival as opposed to the situation where only the FcγRIIB inhibitory receptor is engaged triggering a stress response and apoptosis (Ono et al. 1997).

Natural killer cells

Natural killer (NK) cells are involved in the regulation of hematopoiesis, immune surveillance against tumors and natural immunity to infectious agents. NK cytolytic function is tightly regulated by a group of receptors that recognize class I major histocompatibility complex (MHC) molecules on the surface of target cells (Moretta and Moretta 1997; Lanier 1998). Moreover, these NK receptors can detect and kill abnormal cells that have lost expression of MHC class I (Sentman et al. 1995). Therefore low or absent expression of MHC class I may permit cytolytic killing of virally-infected cells or tumor cells by NK cells. NK receptors that bind to MHC I molecules have been identified in both humans and mice. These receptors belong to two diverse families - the Immunoglobulin (Ig) superfamily and the C-type lectin family.

Human NK cells express Killer Inhibitory Receptors (KIR) that belong to the Ig superfamily. KIR are type I transmembrane glycoproteins that possess one or more extracellular Ig domains thus creating different subfamilies (KIR2 and KIR3). These subfamilies are encoded by genes on chromosome 19 in humans (Wagtmann et al. 1995). KIR homologues have been identified in primates (Viliante et al. 1997), but have not been identified in rat or mouse. However, the first mouse NK receptor that has significant structural homology to human KIR is the ITIM-containing gp49B receptor, originally identified in mouse mast cells (Castells et al. 1994).

In mice NK cells express CD94/NKG2 and Ly49 receptors. Both are members of the C-type lectin family that bind MHC class I molecules. Human NK cells also express NKG2/CD94 complexes which belong to the C-type lectin family. CD94/NKG2 receptors are a family of heterodimeric proteins expressed on the surface of NK cells that are comprised of an invariant CD94 polypeptide disulfide linked to either NKG2A, B, C, D or E (Aramburo et al. 1990; Lazettic et al. 1996; Brooks et al. 1997; Carrertero et al. 1997; Cantoni et al. 1998). Ly49 molecules were first discovered in mice and to date this gene family has at least nine highly related members (Ly49A-I) encoded within the NK complex on mouse chromosome 6 (Yokoyama 1998; Brown et al. 1997; McQueen et al. 1998). Similar to KIRs in the human, individual NK cells can express more than one Ly49 receptor (Raulet et al. 1997; Kubota et al. 1999). A Ly49-like gene was cloned and is thought to be a pseudogene that may represent the remnant of a common ancestral gene (Westgarrd et al. 1998). Interestingly, CD94/NKG2 and Ly49 molecules are considered to be a unique group of C-type lectins, known as group V, which are distinctly different in sequence from other C-type lectins. These receptors lack the majority of the Ca^{2+} binding residues conserved among other C-type lectins (Drickamer 1993). It has been suggested that they constitute a novel family of C-type lectins with NK receptor domains (NKD) (Weis et al. 1998).

Inhibitory and activating receptors expressed by NK cells

KIRs, CD94/NKG2 and Ly49s are classified as either activating or inhibitory receptors. Inhibitory receptors (such as KIR2DL1, 2,3 and 4, KIR3L1 and 2, NKG2A and B, Ly49 A, B, C, E, F, G2 and I) have cytoplasmic ITIMS. The activating receptors (KIR2DS1, 2, 3, 4 and 5, KIR3DS1, NKG2C, NKG2D, NKG2E, LY49D, H, K and N) possess a positively charged residue within the cytoplasmic domain or ITAM motifs. These motifs promote the association of adaptor molecules (such as DAP10 or DAP12) that activate positive signaling cascades (Lanier et al. 1998; Smith et. al. 1998).

Signal transduction pathways of activating NK receptors

Tyrosine phosphorylation of the ITAM in the cytoplasmic domain of activating KIR (KIR2DS1, 2 and 3) receptors by their HLA ligands can induce the association of DAP12 or protein tyrosine phosphatases SHP-1 and SHP-2 both *in vitro* and *in vivo* (Bruhns et al. 1999; Olcese et al. 1996). DAP 12, also known as the Killer cell-Activating-Receptor-Associated Protein (KARAP), associates with activating KIRs (Olcese et al. 1996 and Tomasello et al. 1998). DAP12 has also been shown to associate with Ly49D, Ly49H and CD94/NKG2C (Mason et al. 1998; Lanier et al. 1998; Smith et al. 1998; Cambella and Colonna 1999). DAP10 also associates with NKG2D to form an activating receptor complex (Wu et al. 1999). DAP10 contains a docking site for the SH2 domains of the p85 regulatory subunit of PI3K suggesting that inositol phospholipid signalling pathways may be critical for activation of NK cells (Wu et al. 1999). The NKG2D/DAP-10 complex also recognizes the MICA antigen expressed on tumor cells providing further support for NK cells in immunosurveillance toward tumor cells (Wu et al. 1999).

A role for SHIP in NK cell biology

It has been shown that SHIP represses or limits the activation of immune cells through inhibitory receptors. PI3K has recently been shown to play a role in NK cell activation suggesting that inositol phospholipid signaling presumably has a role in NK cell physiology and function (Wu et al. 1999). Therefore, a role for SHIP in modulating signal transduction pathways active in NK seems likely. Our laboratory has recently generated SHIP$^{-/-}$ mice using a Cre-lox strategy. Analysis of these mice indicates that SHIP plays a role in controlling the expansion of NK cell subsets that express multiple inhibitory receptors specific for self-MHC class I ligands (Howson et al. submitted). We find that NK cells which co-express Ly49A and Ly49C expand such that in adult mice homeostasis of the NK cell compartment is lost. This results in a profound expansion of Ly49A$^+$C$^+$ NK cells.

This reperoire imbalance has functional consequences as well, since adult NK cells, but not cells from juvenile mice, are severely defective in their ability to lyse tumor cells. In summary, our results indicate a novel function for SHIP in the maintenance of the NK cell repertoire and thus the function of NK cells.

There are three potential molecular explanations for the role of SHIP in the dynamics of NK cell subsets (see Fig. 1). Following co-engagement of different Ly49 receptors specific for self MHC ligands SHIP is recruited to the membrane and activated leading to: (1) hydrolysis of PI(3,4,5)P$_3$ that blocks NK cell survival by preventing recruitment and activation of Akt/PKB, (2) SHIP binds to Shc preventing Grb2-mSOS activation of proliferation through the Ras/MAPK pathway or (3) SHIP blocks the activation of transcription factors necessary for the sequential expression of other Ly49 receptors. One or all of these mechanisms could account for the role of SHIP *in vivo* by preventing the NK cell acquisition of multiple inhibitory receptors specific for self MHC ligands.

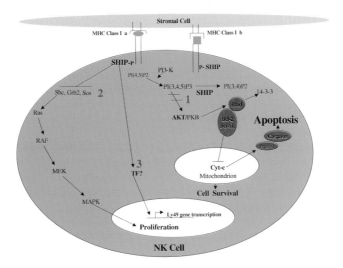

Fig. 1. Potential roles for SHIP in NK cell signaling that could account for the increased number of NK cells in SHIP$^{-/-}$ mice with multiple receptors specific for self MHC ligands. Cross-linking of diverse NK receptors to different self MHC class I molecules leads to binding of SHIP to their ITIM motifs and: (**1**) reduction of basal PI(3,4,5)P$_3$ levels, inhibition of AKT/PKB activation, increased numbers of BAD/Bcl-xL complexes that initiate cytochrome C (Cyt-c) release from the mitochondria leading to cell death, (**2**) SHIP is tyrosine phosphorylated and forms a bidentate interaction with Shc that prevents membrane recruitment of Grb2/mSOS and activation of the Ras/MAPK kinase pathway necessary for the proliferation of NK cells or (**3**) SHIP signaling blocks the activation of transcription factor(s) that induce transcription of additional Ly49 receptors.

A role for SHIP in stem cell biology

We have recently identified a novel SHIP isoform, s-SHIP, whose expression is limited to embryonic and hematopoietic stem cells (Tu et al. 1999; Tu et al., submitted). This isoform is transcribed from an internal promoter and lacks the SH2 domain. It differs from the isoforms expressed in more differentiated cells, since it constitutively associates with Grb2 and is not tyrosine phosphorylated following stimulation of stem cells by LIF or SCF. The highly restricted expression pattern of s-SHIP strongly suggests a specific role for SHIP in signaling pathways critical to the biology of totipotent and pluripotent stem cell populations.

WGK is supported by grants from the NIH (RO1 DK54767, R21 AI44333, PO1 NS27405).

References

Aramburu J, Balboa MA, Ramirez A, Silva A, Acevedo A, Sanchez-Madrid F, De Landazuri MO, Lopez-Botet M (1990) A novel functional cell surface dimer (Kp43) expressed by natural killer cells and T cell receptor-g/d + T lymphocytes. I. Inhibition of the IL-2 dependent proliferation by anti-Kp43 monoclonal antibody. J Immunol 144:3238-3247

Bollard S, Ravetch JV (1998) Inhibitory pathways triggered by ITIM-containing receptors. Adv Immunol 72:149-177

Brooks AG, Posch PE, Scorzelli CJ., Borrego F, Coligan JE (1997) NKG2A complexed with CD94 defines a novel inhibitory NK cell receptor. J Exp Med 185:795-800

Brown MG, Fulmek S, Matsumoto K, Cho R, Lyons PA, Levy ER, Scalzo AA, Yokoyama WM (1997) A 2-Mb YAC containing and physical map of the natural killer gene complex on mouse chromosome 6. Genomics 42:16-25

Bruhns CX, Pesesse C, Moreau D, Blero Erneux C (1999) The two SH2 domain containing inositol 5- phaosphatases SHIP1 and SHIP2 are coexpressed in human T lymphocytes. Biol Chem 380:969-974

Campbell KS, Colonna M (1999) A key accessory protein for relaying signals by natural killer cell receptors. Int J Biochem Cell Biol 31:631-636

Cantoni C, Biassoni R, Pende D, Sivori S, Accame L, Pareti I, Semenzato G, Morretta L, Moretta A, Bottino C (1998) The activating form of CD94 receptor complex: CD94 covalently associates with the Kp39 protein that represents the product of the NKG2-C gene. Eur J Immunol 28:327-338

Carretero M, Cantoni C., Bellon T., Bottino C., Biassoni R., Rodriguez A, Perez-Villar JJ, Moretta L., Moretta A, Lopez-Botet M (1997) The CD94 and NKG2-A C type lectins covalently assemble to form a natural killer cell inhibitory receptor for HLA class I molecules. Eur J Immunol 27:563-575

Castells MC, Wu CX, Arm JP, Austen KF, Katz HR (1994) Cloning of the gp49B gene of the immunoglobulin superfamily and demostration that one of its two products is an early expressed mast cell surface protein originally described as gp49. J Biol Chem 269:8393-8401

Damen JE, Liu L., Cutler RL, Krystal G (1993) Erythropoitetin stimulates the tyrosine phosphorylation of Shc and its association with Grb2 and a 145-Kd tyrosine phosphorylated protein. Blood 82:2296-2303

Damen JE., Liu L., Rosten P, Humphries RK, Jefferson AB, Majerus PW, Krystal G (1996) The 145-kDA protein induced to associate with Shc by multiple cyokines is an inositol tetraaphosphatase ans phosphatidylinositol 3,4,5-trisphosphate 5-phosphatase. Proc Natl Acad Sci USA 93:1689-1693

Damen JE, Liu L, Ware MD, Ermolaeva M, Majerus PW, Krystal G (1998) Multiple forms of the SH2-containing inositol phosphatase. SHIP, are generated by C-terminus truncation. Blood 92:1199-1205

DeFranco AL (1997) The complexity of signals pathways activated by the BCR. Curr Opin Immunol 9:296-308

Drickamer K (1993) Recognition of complex carbohydrates by Ca(2+)-dependent animal lectins. Biochem Soc Trans 2:456-459

Franke TF, Kapal DR, Cantley LC (1997) PI3'K: Downstream AKTion blocks apoptosis. Cell 88:435-437

Hemmings BA (1997) Signal transduction Akt-signaling: Linking membrane events to life and death decision. Science 275:628-630

Howson, JM, Wang, JW, Ghansah T, Ninos JM, Wei S, Loughran TP, Sorimachi N, Kerr WG (2001) The SH2-containing inositol phosphatase limits the expansion of natural killer cells that express inhibitory receptors specific for self MHC ligands. *Submitted.*

Huber M, Helgason CD, Damen JE, Scheid M, Durino V, Liu L, Ware MD, Humphries RK, Krystal G (1999) The role of SHIP in growth factor induced signalling. Progress Biophys Mol Biol (Pawson Aj.ed) pp 423-434. Elsevier Science. The Nertherlands

Katz HR, Viver EE, Castells MC, McCormick MJ, Chambers JM, Austen KF (1996) Proc Natl Acad Sci USA 93 :10809-10814

Kavanaugh WM, Pot DA, Chin SM, Deuterreinhard M, Jefferson AB, Norris FA, Masiarz FR, Cousens LS, Majerus PW, Williams LT (1996) Multiple forms of an inositol polyphosphate 5-phosphatae form signaling complexes with Shc and Grb2. Curr Biol 6:438-445

Klippel A, Kavanaugh WM, Pot D, Williams LT (1997) A specific product of phosphotidylinositol 3-kinase directly acitvates the protein kinse Akt through its pleckstrin homology domain. Mol Cell Biol 17:338-344

Kubota A, Kubota S, Lohwasser S, Mager DL, Takei F (1999) Diversity of NK cell receptor repertoire in adult and neonatal mice. J Immunol 163:212-216

Kuroiwas A, Yamashita Y, InuiM, Yuasa T, Ono M, Nagabukuro A, Matsuda Y, Takai T (1998) Asssociation of tyrosine phosphatases SHP-1 and SHP-2, inositol 5-phosphatase SHIP with gp49B1, and chromosomal assignment of the gene. J Biol Chem 273:1070-1074

Krystal G (2000) Lipid phosphatases in the immune system. Semin Immunol 12:397-403

Lanier LL (1998) NK cell receptors. Annu Rev Immunol 16:359-393

Lanier LL, Corliss BC, Wu J, Leong C, Phillips JH (1998) Association of DAP12 with activating CD94/NKG2C NK cell receptors. Immunity 8:693-701

Lazetic S, Chang C, Houchins JP, Lanier LL, Phillips JH (1996) Human NK cell receptors involved in MHC class I recognition and disulfide-linked heterodimers of CD94 and NKG2 subunits. J Immunol 157:4741-4745

Leibson PJ (1997) Signal transduction during natural killer cell activation: inside the mind of a killer. Immunity 6:655-661

138

Liu L, Damen JE, Culter RL, Krystal G (1994) Multiple cytokines stimulate the binding of a common 145-kilodalton protein to Shc at the Grb2 recognition site of Shc. Mol Cell Biol 14:6926-6935

Lioubin MN, Myles GM, Carlberg K, Bootwell D, Rohrschneider LR (1994) Shc. Grb2, Sos1 and a 150-kilodalton tyrosine-phosphorylated protein form complexes with FMs in hematopoietic cells. Mol Cell Biol 14:5682-5691

Lioubin MN, Algate PA, Tsai S, Carlberg K, Aebersold R, Rohrschneider LR (1996) P150 SHIP, a signal transduction molecule with inositol polyphosphate-5-phosphatase acitivity. Genes Dev 10:1084-1095

Lucas DM, Rohrschneider LR (1999) A novel spliced form of SH2-containing insitol phosphatase SHIP is expressed during myeloid development. Blood 93:1922-1933

Maresco L, Osborne JM, Cooney D, Coggeshall KM, Anderson CL (1999) The SH2-containing 5'-inositol phosphatase (SHIP) is tyrosine phosphorylated after Fc gamma receptor clustering in monocytes. J Immunol 162:6458-6465

Mason LH, Willette-Brown J, Anderson SK, Gosselin P, Shores EW, Love PE, Ortaldo JR, McVicar DW (1998) Characterization of an associated 16-kDa tyrosine phosphoprotein required for ly-49D signal transdcution. J Immunol 160:4148-4152

McQueen K, Freeman JD, Takei F, Mager DL (1998) Localization of five new Ly49 genes, including three closely related to Ly49C. Immunogenetics 48:174-183

Moretta A., Moretta L (1997) HLA class I specfic inhibitory receptors. Curr Opin Immunol 9:694-701

Olcese L, Lang P, Vely F, Cambiaggi A, Marguet D, Blery M, Hippen KL, Biassoni R, Moretta A, Moretta L, Cambier JL, Vivier E (1996) Human and mouse killer cell inhibitory receptors recruit PTP1C and PTP1D protein tyrosine phosphatase. J Immunol 156:4531-4534

Ono M, Bolland S, Tempst P, Ravetch JV (1996) Role of the inositol phosphatase SHIP in negative regulation of the immune system by the receptor FcγRIIB. Nature 383:263-266

Ono M, Okada H, Bolland S, Yanagi S, Kurosaki T, Ravertch JV (1997) Deletion of SHIP or SHP-1 reveals two distinct pathways for inhibitory signaling. Cell 90:293-301

Osborne MA, Zenner G, Lubinus M, Zhang X, Songyang Z., Cantley LC, Majerus P, Burn P, Kochan JP (1996) The insoitol 5'-phosphatase SHIP binds to immunoreceptor signaling motifs and responds to high affinity IgE receptor aggregation. J Biol Chem 271:292971-29278

Pesess XS, Deleu F, DeSmedt L, Drayer, Erneux C (1997) Identification of a second SH2-domain-containg protein closely related to the phosphatidylinositol polyphosphate 5-phosphate SHIP. Biochem Biophys Res Comm 239:697-700

Phillips NE, Parker D C (1984) Cross-linking of B lymphocyte Fc gamma receptors and membrane immunoglobulin inhibits anti-immunoglobulin-induced blastogenesis. J Immunol 132:627-632

Rameh LE, Cantley LC (1999) The role of phosphoinositide 3-kinase lipid products in cell function. J Biol Chem 274:8347-8350

Raulet DH, Held W, Correa I., Dorfman JR, Wu MF, Corall L (1997) Specificity, tolerance and developmental regulation of natural killer cells defined by expression of class I-specfic Ly49 receptors. Immunol Rev 155:41-52

Ravichandran KS, Lee KK, Songyang Z, Cantley LC, Burn P, Burakoff SJ (1993) Interaction of Shc with the zeta chain of the T cell receptor upon T cell activation. Science 262:902-905

Rohrschneider LR, Fuller JF, Wolf I, Liu Y, Lucas DM (2000) Structure, function, and biology of SHIP proteins. Genes Dev 14:505-520

Saxton, TM, Van Ostveen I, Botwell D, Aebersold R, Gold MR (1994) B cell antigen receptor cross-linking induces tyrosine phosphorylation of the p21ras oncoprotein acitvators Shc and Sos1 as well as assembly of complexes containing Shc, Grb2,mSos1, and a 145-KD tyrosine phophorylated protein. J Immunol 153:623-636

Scharenberg AM, Kinet JP (1996) The emerging field of receptor-mediated inhibitory signaling: SHP or SHIP? Cell 87:961-964

Sentman C L, Olsson MY, Karre K (1995) Missing self recognition by natural killer cells in MHC class I transgenic mice. A 'receptor calibration' model for how effector cells adapt to self. Sem Immunol 7:109-119

Smith KM, Wu J, Bakker ABH, Phillips JH, Lanier LL (1998) Ly-49D and Ly-49H associate with mouse DAP12 and form activation receptors. J Immunol 161:7-10

Tomasello E, Olcese L, Vely F, Geourgeion C, Blery M, Moqrich A. Gautheret D, Djabali M, Mattei MG, Viver E (1998) Gene Structure, expression pattern, and biological activity of mouse killer celll activating receptor-associated protein (KARAP)/DAP12). J Biol Chem 273:34115-34119

Tridandapani S, Kelly T, Cooney D, Pradhan M, Coggeshall KM (1997) Negative signaling in B cells: SHIP Grbs Shc. Immunol Today 18:424-427

Tu Z, Ninos J, Wang JW, Lemos M, Kerr WG (1999) Alternative transcriptional initiation in the SHIP gene generates an embryonic isoform that forms a signaling complex which differs from its counterpart expressed in adult hematopoiesis. GenBank, accession number: AF184912

Tu. Z, Ninos J, Ma Z, Lemos M, Wang JW, Kerr WG (2001) Identification of a novel SHIP isoform expressed in embryonic and hematopoietic stem cells that associates with the Grb2 adapter molecule. Submitted

Vance RE, Kraft JR, Altman JD, Jensen PE, Raulet DH (1998) Mouse CD94/NKG2A is a natural killer cell receptor for the nonclassical major histocompatibility complex (MHC) class I molecule Qa-1(b). J Exp Med 188:1841-1848

Valiante NM, Uhrberg M, Shiliing HG, Lienert-Weidenbach K, Arnett KL, D'Andrea A, Philips JH, Lanier LL, Parham P (1997) Functionally and structurally distinct NK cell receptor repertoires in the peripheral blood of two human donors. Immunity 7:739-751

Vely FS, Olcese L, Moretta A, Damen JE, Liu L. Krystal G, Cambier JC, Daeron M, Vivier E (1997) Differential association of phosphatase with Hematopoietic co-receptors bearing immunoreceptor tyrosine-based inhibition motifs. Eur J Immunol 27:1994-2000

Vivier E, Daeron M (1997) Immunoreceptor tyrosine-based inhibition motifs (ITIMs). Immunol Today 18:286-291

Wang, LL, Mehta IKK, LeBlanc PA, Yokoyama WM (1997) Mouse natural killer cells express gp49B1, a structural homologue of human killer inhibitory receptors. J Immunol 158:13-17

Wagtmann N, Biassoni R, Cantoni C, Verdiani S, Malati MS, Vitale M, Bottino C, Morrettal, Moretta A,Long EO (1995) Molecular clones of the p58 NK cell receptor reveal immunoglobulin-related molecules with diversity in both the extra- and intracellular domains. Immunity 2:439-449

Weis WI, Taylor ME, Drickamer K (1998) The C-type lectin superfamily in the immune system. Immunol Rev 163:19-34

Westgaard IH, Berg SF, Orstavik S, Fossum S, Dissen E (1998) Identification of human member of the Ly-49 multigene family. Eur J Immunol 28:1839-1846

Wisniewski D, Strife A, Swendeman S, Erdjument-Bromage H, Geromanos S, Kavanaugh WM, Tempst P, Clarkson B (1999) A novel Sh2 containing phosphtidylinositol 3,4,5-triphosphate 5-phosphatase (SHIP2) is constitutively tyrosine phosphorylated and associated with src homolgous and collagen gene (SHC) in chronic myelogenous leukeima progenitor cells. Blood 93:2707-2720

Wu J, Song Y, Baker AB, Bauer S. Spies T, Lanier LL, Phillips JH (1999) An activating immunoreceptor complex formed by NKG2D and DAP10. Science 285:730-732

Yokoyama WM (1998) Natural Killer Cell Receptors. Curr Opin Immunol 10:298-305

18
SHIP1-mediated negative regulation of cell activation and proliferation by FcγRIIB

Marc Daëron, Pierre Bruhns, Renaud Lesourne, Odile Malbec, and
Wolf H. Fridman

Laboratoire d'Immunologie Cellulaire et Clinique, INSERM U.255, Institut Curie, Paris,
France

Summary. The intracytoplasmic domain of FcγRIIB, a family of widely
expressed IgG receptors, contains an Immunoreceptor Tyrosine-based Inhibition
Motif (ITIM) that accounts for the ability of these receptors to negatively regulate
cell activation and cell proliferation induced by receptors containing
Immunoreceptor Tyrosine-based Activation Motif (ITAMs) and by receptors with
an intrinsic protein tyrosine kinase activity. These properties lie in the affinity of
the FcγRIIB ITIM for SH2 domain-containing inositol 5-phosphatases (SHIPs).
FcγRIIB are the only known ITIM-bearing receptors capable of recruiting SHIPs
in vivo. The present review discusses recent findings that established the
molecular bases for the affinity of the FcγRIIB ITIM for SHIPs, that explained
why SHIPs, rather than SH2 domain-containing protein tyrosine phosphatases, are
recruited by FcγRIIB in vivo, that unraveled a dual mechanism used by FcγRIIB
to negatively regulate B cell activation *via* SHIP1, and that demonstrated the role
of SHIP1 in FcγRIIB-dependent negative regulation of mast cell proliferation
induced by the growth factor receptor Kit.

Key words. Mast cells, B cells, Fc receptors, ITIM, Phosphatases

Introduction

FcγRIIB are a family of single-chain low-affinity receptors for the Fc portion of
IgG antibodies which bind immune complexes with high avidity. They are
encoded by a single gene whose transcripts are alternatively spliced, generating
two (FcγRIIB1 and B2) or three (FcγRIIB1, B1' and B2) isoforms, in humans and
mice respectively. FcγRIIB isoforms are differentially distributed in cells of the
lymphoid and myeloid lineages and, altogether, they are widely expressed, mostly
by cells of hematopoietic origin. FcγRIIB can mediate the internalization of

immune complexes and promote the presentation of antigens, but they cannot activate cells when aggregated at the cell surface. Instead, FcγRIIB inhibit cell activation induced by other receptors (Daëron 1997).

The inhibitory properties of FcγRIIB were first observed *in vivo* when passively administered IgG antibodies to an antigen were found to profoundly suppress antibody responses to that antigen (Henry and Jerne, 1968). As suppression was also induced by antibodies to immunoglobulins and required IgG antibodies with an intact Fc portion, it was proposed that the coaggregation of Fc Receptors for IgG with surface immunoglobulins expressed on the same B cells could inhibit B cell activation (Sinclair and Chan 1971). Surface immunoglobulins were later understood as being the antigen-binding moiety of B cell receptors (BCR), and FcγRIIB were shown as being the only type of FcγR expressed by murine B cells. FcγRIIB-mediated negative regulation was subsequently extended to cell activation triggered by all receptors that bear Immunoreceptor Tyrosine-based Activation Motifs (ITAMs), i.e. not only BCR, but also T cell receptors for antigens (TCR) (Daëron et al. 1995a) and cell-activating Fc Receptors such as high-affinity receptors for IgE (FcεRI) (Daëron et al. 1995b). These in vitro findings were confirmed *in vivo* in several experimental models using mice whose FcγRIIB-encoding gene had been inactivated. Thus, FcγRIIB-deficient mice were found 1) to exhibit enhanced antibody responses (Takai et al. 1996), 2) to develop exaggerated IgE- (Ujike et al. 1999) and IgG-dependent anaphylactic reactions (Takai et al. 1996), 3) to have an enhanced susceptibility to experimental murine models of IgG-dependent autoimmune diseases (Bolland and Ravetch 2000; Clynes et al. 1999; Nakamura et al. 2000; Yuasa et al. 1999), and 4) to exhibit enhanced antibody-dependent cell-mediated cytotoxic responses to the injection of therapeutic antibodies to tumor antigens (Clynes et al. 2000). Recently, FcγRIIB-mediated negative regulation was further extended to growth factor receptors with an intrinsic protein tyrosine kinase activity such as the Stem Cell Factor Receptor Kit that controls the proliferation of mast cells (Malbec et al. 1999).

The inhibitory properties of FcγRIIB were shown to reside in a 13 aminoacid intracytoplasmic sequence that is conserved in all mouse, rat and human FcγRIIB isoforms (Amigorena et al. 1992; Muta et al. 1994). The same 13-aminoacid sequence was required for inhibition of cell activation, irrespectively of the type of ITAM-bearing receptor, in B cells, T cells and mast cells (Daëron et al. 1995a). In the 13-aminoacid inhibitory sequence of FcγRIIB, a tyrosine residue, followed at position Y+3 by a leucine residue, was shown to be mandatory for inhibition (Muta et al. 1994; Daëron et al. 1995a). One or several copies of similar sequences were subsequently found in the intracytoplasmic domains of a large number of membrane molecules that could all inhibit cell activation (Daëron and Vivier 1999). Sequence comparison between these inhibitory molecules permitted to define an Immunoreceptor Tyrosine-based Inhibition Motif (ITIM) consisting of a tyrosine followed by a leucine or valine, at position Y+3, and preceded by a loosely conserved aminoacid (mostly isoleucine, leucine or valine, sometimes serine or threonine), at position Y-2 (Burshtyn et al. 1997; Vivier and Daëron

1997). ITIMs provided the structural grounds for a general scenario capable of explaining how ITIM-bearing molecules can negatively regulate cell activation. When inhibitory receptors are coaggregated with activating receptors, their ITIMs are the substrates of a Protein Tyrosine Kinase (PTK) of the src family (Malbec et al. 1998). Tyrosyl-phosphorylated ITIMs subsequently recruit SH2 domain-containing phosphatases that interfere with signal transduction. With one exception, all ITIM-bearing receptors studied were found to recruit the protein tyrosine phosphatases SHP-1 and/or SHP-2 (collectively referred to as SHPs). SHPs dephosphorylate tyrosines in receptors and in effector molecules whose tyrosyl-phosphorylation is critical for activation signals to be transduced. As a consequence of this inhibition of early signaling events, downstream signals are not generated and the cell remains non-activated (Long, 1999). FcγRIIB are the exception. Instead of protein tyrosine phosphatases, tyrosyl-phosphorylated FcγRIIB recruit the SH2 domain-containing inositol 5-phosphatases SHIP1 (Ono et al. 1996; Fong et al. 1996) and SHIP2 (Muraille et al., 2000) (collectively referred to as SHIPs). The preferred substrate of SHIPs is phosphatidylinositol (3,4,5)-trisphosphate (PI(3,4,5)P3) which enables the membrane recruitment of several molecules having a Pleckstrin Homology (PH) domain. Among these is the Bruton's Tyrosine Kinase (BTK) which is mandatory for Phospholipase C-γ (PLC-γ) to be activated and to induce a Ca^{2+} response (Bolland et al. 1998; Scharenberg et al., 1998). By recruiting SHIPs, FcγRIIB therefore do not prevent early activation signals but arrest the intracellular propagation of downstream signals.

Works discussed below aimed at further understanding SHIP-mediated inhibitory mechanisms used by FcγRIIB and more specifically 1) at defining the molecular basis of the unique affinity of the FcγRIIB ITIM for SHIPs, 2) at understanding the molecular bases of the *in vivo* preference of FcγRIIB for SHIPs, 3) at examining the mechanisms of SHIP-mediated inhibition, and 4) at evaluating the role of SHIPs in FcγRIIB-mediated negative regulation of RTK-dependent cell proliferation.

Molecular basis of the affinity of FcγRIIB for SHIPs

In order to identify aminoacids that account for the specificity of the FcγRIIB ITIM for SHIPs, these were exchanged for corresponding aminoacids that are present in the N-terminal ITIM of a human killer cell Ig-like inhibitory receptor called KIR2DL3 (KIR N-ITIM). KIR2DL3 ITIMs had been previously demonstrated to bind SHP-1 and SHP-2, but not SHIP1 (Vély et al. 1997). Within the 13-aminoacid sequence that was originally shown to be necessary for the inhibitory properties of murine FcγRIIB, the FcγRIIB ITIM can be defined as AENT**ITY**SL**L**KHP (in bold face). The point mutation of the tyrosine abolished the ability of receptors to recruit phosphatases (Ono et al. 1996; Malbec et al. 1998) and to inhibit cell activation (Daëron et al., 1995). Y+3 residues were shown to be directly involved in the binding of tyrosyl-phosphorylated sequences to SH2 domains (Waksman et al. 1992). We found previously that both the Y-2 isoleucine of the FcγRIIB ITIM and the Y-2 valine of the KIR N-ITIM are critical for the ability of corresponding phosphopeptides to bind SHP-1 and SHP-2 in vitro (Vély et al. 1997). The Y-1 threonine being conserved in the FcγRIIB ITIM and in the KIR N-ITIM, it could be excluded as being critical for SHIP binding. We therefore exchanged the serine and the leucine that are present at positions Y+1 and Y+2 in FcγRIIB, with the alanine and the glutamine that are present at the same positions in the KIR N-ITIM (Bruhns et al. 2000).

Thus, a $SL_{(+1+2)}AQ$ mutation was made in FcγRIIB1 and, conversely, an $AQ_{(+1+2)}SL$ mutation was made in FcγRIIB1 whose ITIM had been replaced by the KIR N-ITIM. Wild-type and mutant FcγRIIB1 were stably expressed in a mast cell and in a B cell line. When coaggregated with FcεRI or with BCR, FcγRIIB1 bearing a $SL_{(+1+2)}AQ$ mutation became tyrosyl-phosphorylated, but lost the ability to recruit both SHIP1 and SHIP2. Under the same conditions, FcγRIIB1 bearing the KIR N-ITIM, recruited no phosphatase although it was phosphorylated. Remarkably, an $AQ_{(+1+2)}SL$ mutation rendered this chimeric receptor capable of recruiting both SHIP1 and SHIP2. The same aminoacid substitutions were introduced in synthetic peptides having the sequences of the FcγRIIB ITIM or of the KIR N-ITIM, and corresponding phosphorylated peptides were used to precipitate phosphatases from cell lysates. Amino acid substitutions exhibited the same positive and negative effects on the in vitro binding of SHIP1 and SHIP2 as they did on the in vivo recruitment of SHIP1 and SHIP2. They had no effect on the binding of SHP-1 and SHP-2. Identical results were observed when the affinity of these peptides for the GST fusion proteins containing the SH2 domains of phosphatases was examined. Finally, individual $A_{(+1)}S$ and $Q_{(+2)}L$ substitutions, in the FcγRIIB ITIM, and $S_{(+1)}A$ and $L_{(+2)}Q$ substitutions, in the KIR ITIM, identified the Y+2, but not the Y+1 residue, as determining the affinity for both SHIP1 and SHIP2. FcγRIIB is the only ITIM-bearing molecule that contains a leucine at position Y+2. Interestingly, one tyrosine-containing SHIP1-binding sequence was identified in the erythropoietin receptor. It contains an isoleucine at this position.

Altogether, these results indicate that two symmetrical highly hydrophobic aminoacids, at position Y-2 and Y+2, determine the affinity of ITIMs for SHPs and for SHIPs, respectively. These residues define ITYsLL and ITYsLL as two phosphatase-binding sites, for SHPs and SHIPs respectively, in the FcγRIIB ITIM (Bruhns et al. 2000). This having been established, an intriguing question was why the SHIP-binding site is preferentially used in vivo by FcγRIIB, although the two binding sites seem to function equally well in vitro.

Molecular basis of the selective recruitment of SHIPs by FcγRIIB in vivo

The nature of phosphatases that are recruited by FcγRIIB in vivo has been a matter of controversy since the pioneer observation by d'Ambrosio et al. was published in 1995. These authors found that SHP-1 coprecipitated with FcγRIIB, when tyrosyl-phosphorylated upon coaggregation with BCR in A20 and in IIA1.6 B cells reconstituted with FcγRIIB, and that FcγRIIB-dependent inhibition of B cell proliferation was impaired in B cells from SHP-1-deficient motheaten mice (D'Ambrosio et al. 1995). In 1996, Ono et al. reported that SHIP1 coprecipitated with FcγRIIB following coaggregation with FcεRI in bone marrow-derived mast cells (BMMCs) or with BCR in A20 cells, and that FcγRIIB-dependent inhibition of IgE-induced serotonin release was unaffected in BMMCs derived from motheaten mice (Ono et al. 1996). The latter observations were confirmed by Fong et al. who reported that SHIP1, but not SHP-1 or SHP-2, coprecipitated with FcγRIIB following coaggregation with FcεRI in BMMCs (Fong et al. 1996). In 1997, Ono et al. showed that FcγRIIB-dependent inhibition of Ca^{2+} responses and of NF-AT activity was abolished in SHIP1-deficient, but not in SHP-1-deficient, DT40 chicken B cells, and that SHIP1, but not SHP-1, was detectably coprecipitated with FcγRIIB following coaggregation with BCR in A20 cells (Ono et al. 1997). The same year, Nadler et al. confirmed that SHP-1 was dispensable for FcγRIIB-dependent inhibition of BCR signaling in B cell lines immortalized from motheaten mice (Nadler et al. 1997). In 1998, however, Sato et al. observed the coprecipitation of both SHIP1 and SHP-1 with FcγRIIB in A20 cells transfected with an anti-TNP BCR, following coaggregation with intact anti-idiotypic antibodies (Sato and Ochi 1998). Contrasting with the consensus that FcγRIIB recruit SHIP1 both in B cells and in mast cells, their ability to recruit SHP-1 in vivo therefore remained controversial.

In an attempt to clarify this issue, we used a variety of ligands to coaggregate FcγRIIB with BCR in B cells or with FcεRI in mast cells, including high concentrations of preformed immune complexes made of multivalent antigens and IgG antibodies in various proportions. These induced a dose-dependent tyrosyl-phosphorylation of FcγRIIB in the two cells types, but even when using optimal concentration of complexes (that induced a maximal phosphorylation of FcγRIIB), SHIP1, but not SHP-1, coprecipitated with the receptors. SHP-1 coprecipitated however when, and only when FcγRIIB were hyperphosphorylated following the treatment of cells with pervanadate. The simplest explanation for this effect of pervanadate was that the recruitment of SHP-1 might require a higher level of FcγRIIB phosphorylation than the recruitment of SHIP1. To examine this possibility, we studied the in vitro binding of SHP-1 and SHIP1 to agarose beads coated with a mixture of non-phosphorylated and phosphorylated FcγRIIB ITIM peptides, so that the total amount of peptides remained constant but the proportion of phosphorylated and non-phosphorylated peptides varied. We found that the binding of SHP-1 decreased more sharply with the proportion of phosphorylated

ITIM present on the beads than the binding of SHIP1, and a selective binding of SHIP1 was observed for proportions of phosphorylated ITIM lower than 12%. Identical results were obtained using beads coated with GST fusion proteins containing either the 13-aminoacid inhibitory sequence or the whole intracytoplasmic domain of FcγRIIB that had been phosphorylated for various periods of time in vitro by the Src ·kinase Lyn. Using a constant amount of phosphorylated ITIM to coat beads in variable numbers, we found that the density of phosphorylated ITIM dramatically affected the binding of SHP-1, but not the binding of SHIP1. We observed also that GST fusion proteins containing the two SH2 domains of SHP-1, but not fusion proteins containing one SHP-1 SH2 domain only, bound to phosphorylated ITIM-coated beads as efficiently as the single SH2 domain of SHIP1, provided the density of peptides was high enough (Lesourne et al. 2001).

These results altogether indicate that the binding of SHP-1 to the phosphorylated FcγRIIB ITIM requires the cooperative binding of the two SH2 domains of the phosphatase and that, for this reason, the recruitment of SHP-1 requires that a higher proportion of FcγRIIB be phosphorylated than the recruitment of SHIP1. Whether such a high enough proportion can be reached under physiological (or pathological?) conditions remains to be determined. If so, FcγRIIB could use two mechanisms to negatively regulate cell activation, with different consequences, and thereby adjust the intensity of the regulation to the intensity of activation stimuli.

Dual mechanism of inhibition by SHIP1 in B cells

Whatever the answer to the above question, all data concur to indicate that, under mild (physiological?) stimulation conditions, FcγRIIB use preferentially SHIPs, if not exclusively, to negatively regulate both BCR-dependent B cell activation and FcεRI-dependent mast cell activation. In the two cell types, the inhibition of biological responses can be accounted for by the extinction of the Ca^{2+} response, resulting from the degradation of PI(3,4,5)P3 by SHIPs (Scharenberg et al. 1998) and the subsequent release of BTK from the membrane (Bolland et al. 1998). Ca^{2+} responses are necessary not only for the exocytosis of mast cell granules that contain preformed inflammatory mediators, but also for NF-AT to be translocated to the nucleus and to combine with transcription factors of the AP-1 complex for inducing the transcription of cytokine genes. Indeed, we found that a chimeric molecule made of FcγRIIB whose intracytoplasmic domain had been replaced by the catalytic domain of SHIP1, exerted the same effects as FcγRIIB1, not only on Ca^{2+} responses, but also on late secretory responses of both mast cells and B cells (Hardré-Liénard et al., submitted).

Inhibition of the Ca^{2+} response, however, does not explain satisfactorily that FcγRIIB can also inhibit the activation of Erk1 and Erk2 (Sarmay et al. 1996), the effector MAP kinases of the Ras pathway. To explain this property of FcγRIIB, it had been proposed that SHIP1, which was shown to form complexes with Shc and Grb2, could sequester these adapter molecules that connect immunoreceptors with Ras, thus preventing downstream events from being triggered (Tridandapani et al. 1997). Another mechanism by which SHIP1 inhibits the activation of Erk1/2 in B cells was recently unraveled by the group of John Cambier. This finding originated from the observation that the basal tyrosyl-phosphorylation of both the adapter molecule p62dok and SHIP1 markedly increased following the coaggregation of FcγRIIB with BCR, and that phosphorylated p62dok coprecipitated with rasGAP. RasGAP stimulates the auto-GTPase activity of Ras and therefore antagonizes with Sos to activate this small G protein. Remarkably, the inducible phosphorylation of p62dok was lost in B cells from SHIP1$^{-/-}$ mice, suggesting that p62dok might bind to SHIP, following the recruitment of the latter by FcγRIIB and its tyrosyl-phosphorylation. Such an adapter function of SHIP1 was demonstrated using chimeric molecules made of FcγRIIB whose intracytoplasmic domain had been replaced either by the PTB domain-containing N-terminal half of p62dok (N-1/2 dok) or by the tyrosine-rich C-terminal half of p62dok (C-1/2 dok). The C-1/2 dok chimera, but not the N-1/2 dok chimera, was constitutively tyrosyl-phosphorylated when expressed in B cells. Upon coaggregation with BCR, the phosphorylation of both chimeras increased and SHIP1, but not rasGAP, coprecipitated with the N-1/2 dok chimera, whereas rasGAP, but not SHIP1, coprecipitated with the C-1/2 dok chimera. Moreover, when coaggregated with BCR, the C-1/2 dok chimera, but not the N-1/2 dok chimera, inhibited the activation of Erk1/2 as efficiently as w.t. FcγRIIB1. The inhibitory effects of FcγRIIB on the Ras pathway can therefore be reproduced by a molecule that bypasses SHIP1. It follows that SHIP1 is necessary for recruiting p62dok, but that its enzymatic activity is not mandatory for inhibiting the activation of MAP kinases (Tamir et al. 2000).

Altogether the above data indicate that SHIP1 can have a dual effect during negative regulation of B cell activation by FcγRIIB. It can extinguish the Ca^{2+} response by acting as an inositol phosphatase and abrogate the activation of MAP kinases by acting as an adapter molecule.

SHIP1 in FcγRIIB-dependent inhibition of Kit-mediated mast cell proliferation

As early experiments were demonstrating that IgG antibodies suppressed a secondary in vitro antibody response of spleen B cells (Henry and Jerne 1968), other experiments showed that F(ab')2 fragments of anti-immunoglobulin antibodies could trigger spleen B cells to proliferate, but not intact antibodies unless the binding site of FcγRIIB had been blocked (Phillips and Parker, 1984). This suggested that antibodies might affect both B cell activation and proliferation. Activation and proliferation, however, are triggered by the same receptor (BCR) and closely linked in B cells, and it was conceivable, though not demonstrated, that inhibition of cell proliferation was the mere consequence of inhibition of cell activation. By contrast with B cells, activation and proliferation can be dissociated in mast cells in which FcεRI trigger cell activation, but not proliferation. Conversely, Kit triggers proliferation but no or little activation. Kit is a prototypic growth factor receptor with an intrinsic protein tyrosine kinase activity. When dimerized by Stem Cell Factor, Kit autophosphorylates. Phosphorylated Kit recruits and activates a variety of effector and adapter molecules that stimulate an array of signaling pathways, ultimately leading cells to proliferate. Our recent finding that FcγRIIB could inhibit Kit-mediated proliferation of mast cells derived in vitro from primary cultures of bone marrow cells (Malbec et al. 1999) provided an appropriate model to investigate how FcγRIIB can negatively regulate cell proliferation. We therefore compared signaling events observed following the aggregation of Kit or the coaggregation of Kit with FcγRIIB that are constitutively expressed by mouse mast cells (Benhamou et al. 1990).

Inhibition of thymidine incorporation induced by coaggregating the two receptors was correlated with a decrease in the numbers of cells entering the G1 phase of the cell cycle. The entry and the progression of cells through the cell cycle are under the control of cyclins that are inducibly expressed upon Kit aggregation. We found that the induction of cyclins D2 and D3, which control the entry into G1, and of cyclin A, which controls the progression into the S phase, were inhibited following the coaggregation of Kit with FcγRIIB. Cyclin induction depends on the coordinated activation of MAP kinases of the Ras (Erk1/2) and of the Rac (JNK and p38) pathways. Erk1/2, JNK and p38 activation were decreased following the coaggregation of Kit with FcγRIIB. The Rac pathway is initiated by the exchange factor Vav which is recruited to the membrane *via* PI(3,4,5)P3. Because PI(3,4,5)P3 also determines the phosphorylation of the protein kinase PKB/Akt and its activation, Akt phosphorylation can be considered as reflecting the amount of PI(3,4,5)P3 in the membrane. Akt phosphorylation seen following Kit aggregation was decreased following the coaggregation of Kit with FcγRIIB. Since PI(3,4,5)P3 is degraded by SHIPs, we looked for the recruitment of these phosphatases by FcγRIIB. SHIP1, but not SHIP2, SHP-1 or SHP-2 coprecipitated with FcγRIIB, when phosphorylated upon coaggregation with Kit. The above findings suggesting a prominent role of SHIP1, we examined mast cells derived from the bone marrow of SHIP1[-/-] mice. The inhibition of Akt phosphorylation and Erk activation were abrogated in SHIP1[-/-] cells. Surprisingly, however, the inhibition of cyclin D3 induction was reduced, but not abrogated. Likewise, inhibition of thymidine incorporation was reduced, but only partially. Remarkably, FcγRIIB-dependent inhibition of FcεRI-mediated mast cell activation was abolished in SHIP1[-/-] mast cells. Finally, however, FcγRIIB-dependent inhibition of Kit-mediated mast cell proliferation could be mimicked by FcγRIIB whose intracytoplasmic domain was replaced by the catalytic domain of SHIP1 (Malbec et al., submitted).

It follows that SHIP1 can account for inhibition of Kit-mediated mast cell proliferation. SHIP1-independent mechanisms, however, can either replace SHIP1-dependent mechanisms in SHIP1$^{-/-}$ cells, or complement SHIP1-dependent mechanisms in w.t. mast cells.

Conclusion

A still increasing number of ITIM-bearing inhibitory molecules have been described after an ITIM was first identified in FcγRIIB. FcγRIIB, however, appear to be unique, when compared with these many molecules. The reason lies in unique properties of the FcγRIIB ITIM that enables these receptors to recruit inositol 5-phosphatases with a single SH2 domain, when they are tyrosyl-phosphorylated upon coaggregation with ITAM-bearing receptors. The unique affinity of FcγRIIB for SHIP1 and SHIP2 is indeed determined by a single hydrophobic residue, at position Y+2 within the ITIM core, which is not found in other ITIMs. Like other ITIMs, the FcγRIIB ITIM has nevertheless an affinity for the protein tyrosine phosphatases SHP-1 and SHP-2 that is determined by another hydrophobic residue at position Y-2, but the in vivo recruitment of these molecules with two SH2 domains requires a higher proportion of phosphorylated FcγRIIB than the recruitment of SHIPs. Whether FcγRIIB might switch from SHIP to SHP recruitment when activation signals reach a threshold and whether this could occur, in pathologic situations for instance, are intriguing possibilities. Whether SHIP1 and SHIP2 have redundant or complementary roles is another unanswered question. Whatever the answers, the recruitment of SHIP1, under physiological conditions, can account for most inhibitory properties of FcγRIIB, at least in mast cells and in B cells. SHIP1 indeed inhibits the Ca^{2+} response, by virtue of its 5-phosphatase activity, as well as the activation of Ras, independently of its enzymatic properties, by functioning as an adapter molecule. As a consequence, Ca^{2+}-dependent and MAP kinase-dependent responses triggered by ITAM-bearing receptors are inhibited, and the entry of cells into the cell cycle triggered by protein tyrosine kinase receptors is prevented as a result of an inhibition of the transcription of cyclin genes. In the presence of appropriate IgG antibodies, FcγRIIB may thus coordinately control the development and the activation of a variety of FcγRIIB-expressing hematopoietic cells involved in both the induction and the effector phases of an immune response.

Acknowledgements

Works discussed in the above review were supported by the Institut National de la Santé et de la Recherche Médicale (INSERM), the Institut Curie and the Association pour la Recherche sur le Cancer (ARC). They were conducted in part in collaboration with Dr. Eric Vivier, at the Centre d'Immunologie de Marseille-Luminy, Marseille, France, Dr. John C.

Cambier at the National Jewish Medical and Research Center, Denver, CO, USA., Dr. Christian Schmitt, at the Hôpital de la Pitié-Salpêtrière, Paris, France and Dr. Gerald Krystal at the Terry Fox Laboratory, Vancouver, BC, Canada.

Pierre Bruhns is the recipient of a fellowship from the Société Française du Cancer. Renaud Lesourne is the recipient of a fellowship from the Ministère de l'Enseignement Supérieur et de la Recherche.

References

Amigorena S, Bonnerot C, Drake J, Choquet D, Hunziker W, Guillet JG, Webster P, Sautès C, Mellman I, Fridman WH (1992) Cytoplasmic domain heterogeneity and functions of IgG Fc receptors in B-lymphocytes. Science 256:1808-1812

Benhamou M, Bonnerot C, Fridman WH, Daëron M (1990) Molecular heterogeneity of murine mast cell Fcγ receptors. J Immunol 144:3071-3077

Bolland S, Pearse RN, Kurosaki T, Ravetch JV (1998) SHIP modulates immune receptor responses by regulating membrane association of Btk. Immunity 8:509-516

Bolland S, Ravetch JV (2000) Spontaneous autoimmune disease in FcγRIIB-deficient mice results from strain-specific epistasis. Immunity 13:277-285

Bruhns P, Vély F, Malbec O, Fridman WH, Vivier E, Daëron M (2000) Molecular basis of the recruitment of the SH2 domain-containing inositol 5-phosphatases SHIP1 and SHIP2 by FcγRIIB. J Biol Chem 275:37357-37364

Burshtyn DN, Yang W, Yi T, Long EO (1997) A novel phosphotyrosine motif with a critical amino acid at position -2 for the SH2 domain-mediated activation of the tyrosine phosphatase SHP-1. J Biol Chem 272:13066-13072

Clynes R, Maizes JS, Guinamard R, Ono M, Takai T, Ravetch JV (1999) Modulation of Immune Complex-induced Inflammation In Vivo by the Coordinate Expression of Activation and Inhibitory Fc Receptors. J Exp Med 189:179-185

Clynes RA, Towers LT, Presta LG, Ravetch JV (2000) Inhibitory Fc receptors modulate in vivo cytoxicity against tumor targets. Nature Medicine 6:443-446

D'Ambrosio D, Hippen KH, Minskoff SA, Mellman I, Pani G, Siminovitch KA, Cambier JC (1995) Recruitment and activation of PTP1C in negative regulation of antigen receptor signaling by FcγRIIB1. Science 268:293-296

Daëron M (1997). Fc Receptor Biology. Annu Rev Immunol 15:203-234

Daëron M, Latour S, Malbec O, Espinosa E, Pina P, Pasmans S, Fridman WH (1995a) The same tyrosine-based inhibition motif, in the intracytoplasmic domain of FcγRIIB, regulates negatively BCR-, TCR-, and FcR-dependent cell activation. Immunity 3:635-646

Daëron M, Malbec O, Latour S, Arock M, Fridman WH (1995b) Regulation of high-affinity IgE receptor-mediated mast cell activation by murine low-affinity IgG receptors. J Clin Invest 95:577-585

Daëron M, Vivier E (1999) Biology of Immunoreceptor Tyrosine-based Inhibtion Motif-bearing molecules. Cur. Top. Microbiol Immunol 244:1-12

Fong DC, Malbec O, Arock M, Cambier JC, Fridman WH, Daëron M (1996) Selective in vivo recruitment of the phosphatidylinositol phosphatase SHIP by phosphorylated FcγRIIB during negative regulation of IgE-dependent mouse mast cell activation. Immunol Letters 54:83-91

Henry C, Jerne NK (1968) Competition of 19S and 7S antigen receptors in the regulation of the primary immune response. J Exp Med 128:133-152

Lesourne R, Bruhns P, Fridman WH, Daëron M (2001) Insufficient phosphorylation prevents FcγRIIB from recruiting the SH2 domain-containing protein tyrosine phosphatase SHP-1. J Biol Chem in press

Long EO (1999) Regulation of immune responses through inhibitory receptors. Annu Rev Immunol 17:875–904

Malbec O, Fong D, Turner M, Tybulewicz VLJ, Cambier J, C., Fridman WH, Daëron M (1998) FcεRI-associated lyn-dependent phosphorylation of FcγRIIB during negative regulation of mast cell activation. J Immunol 160:1647-1658

Malbec O, Fridman WH, Daëron M (1999) Negative regulation of c-kit-mediated cell proliferation by FcγRIIB. J Immunol 162:4424-4429

Malbec O, Fridman WH, Daëron M (1999) Negative regulation of hematopoietic cell activation and proliferation by FcγRIIB. Cur Top Microbiol Immunol 244:13-27

Muraille E, Bruhns P, Pesesse X, Daëron M, Erneux C (2000) The SH2 domain containing inositol 5-phosphatase SHIP2 associates to the immunoreceptor tyrosine-based inhibition motif of FcgRIIB in B cells under negative signalling. Immunol Lett 72:7-15

Muta T, Kurosaki T, Misulovin Z, Sanchez M, Nussenzweig MC, Ravetch JV (1994) A 13-amino-acid motif in the cytoplasmic domain of FcγRIIB modulates B-cell receptor signalling. Nature 368:70-73

Nadler MJS, Chen B, Anderson JS, Wortis HH, Neel BG (1997) Protein-tyrosine phosphatase SHP-1 is dispensable for FcγRIIB-mediated inhibition of B cell antigen receptor activation. J Biol Chem 272:20038-20043

Nakamura A, Yuasa T, Ujike A, Ono M, Nukiwa T, Ravetch JV, Takai T (2000) Fcγ receptor IIB-deficient mice develop Goodpasture's syndrome upon immunisation with type IV collagen: A novel murine model for autoimmune glomerular basement membrane disease. J Exp Med 191:899-905

Ono M, Bolland S, Tempst P, Ravetch JV (1996) Role of the inositol phosphatase SHIP in negative regulation of the immune system by the receptor FcγRIIB. Nature 383:263-266

Ono M, Okada H, Bolland S, Yanagi S, Kurosaki T, Ravetch JV (1997) Deletion of SHIP or SHP-1 reveals two distinct pathways for inhibitory signaling. Cell 90:293-301

Phillips NE, Parker DC (1984) Cross-linking of B lymphocyte Fcγ receptors and membrane immunoglobulin inhibits anti-immunoglobulin-induced blastogenesis. J Immunol 132:627-632

Sarmay G, Koncz G, Gergely J (1996) Human type II Fcγ receptors inhibit B cell activation by interacting with the p21ras-dependent pathway. J Biol Chem 271:30499-30504

Sato K, Ochi A (1998) Superclustering of B cell receptor and FcγRIIB activates Src Homology 2-containing protein tyrosine phosphatase-1. J Immunol 161:2716-2722

Scharenberg AM, El-Hillal O, Fruman DA, Beitz LO, Li Z, Lin S, Gout I, Cantley LC, Rawlings DJ, Kinet J-P (1998). Phosphatidylinositol-3,4,5-triphosphate (PtdIns-3,4,5-P3)/Tec kinase-dependent calcium signaling pathway: a target for SHIP-mediated inhibitory signals. EMBO J 17:1961-1972

Sinclair NRSC, Chan PL (1971) Regulation of the immune responses. IV. The role of the Fc-fragment in feedback inhibition by antibody. Adv Exp Med Biol 12:609-615

Takai T, Ono M, Hikida M, Ohmori H, Ravetch JV (1996) Augmented humoral and anaphylactic responses in FcγRII-deficient mice. Nature 379:346-349

Tamir I, Stolpa JC, Helgason CD, Nakamura K, Bruhns P, Daëron M, Cambier JC (2000) The RasGAP-binding protein p62dok is a Mediator of Inhibitory FcγRIIB Signals in B cells. Immunity 12:347-358

Tridandapani S, Chacko GW, Van Brocklyn JR, Coggeshall LM (1997) Negative signaling in B cells causes reduced ras activity by reducing Shc-Grb2 interactions. J Immunol 158:1125-1132

Ujike A, Ishikawa Y, Ono M, Yuasa T, Yoshino T, Fukumoto M, Ravetch J, Takai T (1999) Modulation of immunoglobulin (Ig)E-mediated systemic anaphylaxis by low-affinity Fc receptors for IgG. J Exp Med 189:1573-1579

Vély F, Olivero S, Olcese L, Moretta A, Damen JE, Liu L, Krystal G, Cambier JC, Daëron M, Vivier E (1997) Differential association of phosphatases with hematopoietic coreceptors bearing Immunoreceptor Tyrosine-based Inhibition Motifs. Eur J Immunol 27:1994-2000

Vivier E, Daëron M (1997) Immunoreceptor tyrosine-based inhibition motifs. Immunol. Today 18:286-291

Waksman G, Kominos D, Robertson SC, Pant N, Baltimore D, Birge RB, Cowburn D, Hanafusa H, Mayer BJ, Overduin M, et al. (1992) Crystal structure of the phosphotyrosine recognition domain SH2 of v-src complexed with tyrosine-phosphorylated peptides. Nature 358:646-53

Yuasa T, Kubo S, Yoshino T, Ujike A, Matsumura K, Ono M, Ravetch JV, Takai T (1999) Deletion of Fcγ Receptor IIB Renders H-2b Mice susceptible to Colagen-induced Arthritis. J Exp Med 189:187-194

19
The preBCR signaling through Igβ regulates locus accessibility for ordered immunoglobulin gene rearrangements

Kazushige Maki[1,2], Kisaburo Nagata[1], Fujiko Kitamura[1], Toshitada Takemori[2], and Hajime Karasuyama[1,2]

[1]Department of Immunology, The Tokyo Metropolitan Institute of Medical Science, Bunkyo-ku, Tokyo 113-8613, Japan
[2]Department of Immune Regulation, Tokyo Medical and Dental University, Graduate School, Bunkyo-ku, Tokyo 113-8519, Japan
[3]Department of Immunology, National Institute of Infectious Diseases, Shinjuku-ku, Tokyo 162-8640, Japan

Summary. In B cell development, the rearrangement is initially targeted to the immunoglobulin (Ig) heavy (H) chain locus and then redirected to the light (L) chain locus once μ H chain protein is expressed. To study the role of the preB cell receptor (preBCR) signaling in the regulation of the ordered Ig gene rearrangement during B cell differentiation, a newly developed system using μ H chain membrane exon (μm)-deficient mice was employed. By using this system to manipulate the rearrangement program in vivo, we found that the signaling through Igβ, a component of preB cell receptor, induces the redirection of Ig gene rearrangements, namely, the suppression of rearrangements at the H chain locus while the activation of rearrangements at the L chain locus. Upon the cross-linking of Igβ, the κL chain germline transcription was found to be up-regulated while the V_H germline transcription was down-regulated. Notably, this alteration of the accessibility at the H and L chain loci was detectable even prior to the cellular differentiation indicated by the change of surface phenotype. Thus, the preBCR signaling through Igβ appears to regulate the ordered Ig gene rearrangement by altering the Ig locus accessibility.

Key words. PreB cell receptor, Surrogate light chain, Allelic exclusion, B cell development, Recombinase

The Ig gene rearrangement is tightly regulated in stage-specific fashion during B cell development such that there is a clear-cut order of Ig gene rearrangements, namely, the H chain gene is usually assembled prior to L chain gene rearrangements (Alt et al. 1981; Reth et al. 1985). Expression of μ H chain through a productive $V_H D_H J_H$ rearrangement has been suggested to transduce a signal that inhibits V_H to $D_H J_H$ rearrangement at the other allele of the H chain gene while it facilitates the onset of V-J joining at the L chain locus (Rajewsky 1996).

PreBCR is composed of μH chain, VpreB/λ5 surrogate L chain and Igα/Igβ heterodimer (Karasuyama et al. 1990; Tsubata et al. 1990). The expression of preBCR is restricted to the early preB cell stage in which the rearrangement of the H chain gene but not yet the L chain gene has been completed (Karasuyama et al. 1994; Lassoued et al. 1993). Successful preBCR assembly induces several hallmark events associated with progression from the proB to preB cell stage (Melchers et al. 1993, Karasuyama et al. 1996). However, signals involved in the preBCR function remained largely unknown (Karasuyama et al. 1996). We recently established a novel system to analyze preBCR signaling by using bone marrow proB cells (Nagata et al. 1997). In vivo treatment of RAG-2-deficient mice with anti-Igβ mAb revealed that the cross-linking of Igβ on proB cells induced their differentiation to the small preB cell stage as if it mimics preBCR signaling. Since V(D)J rearrangements shift from the H chain locus to the L chain locus at the transition from the proB to small preB cell stage, we applied this system to study the role of preBCR in Ig gene rearrangement (Maki et al. 2000). For this purpose, μm-deficient mice were used, in which B cell development was completely arrested at the proB cell stage as in RAG-2-deficient mice but the recombination machinery was kept intact in contrast to RAG-2 deficient mice (Shinkai et al. 1992; Kitamura et al. 1991).

μm-deficient mice were treated by a single injection of anti-Igβ mAb HM79 (Koyama et al. 1997). Flow cytometric analysis revealed that the cross-linking of Igβ on proB cells up-regulated the expression of CD25, BP-1 and CD2 while it down-regulated that of c-kit and λ5 (Maki et al. 2000). Therefore, we concluded that the cross-linking of Igβ on proB cells in μm-deficient mice induced the phenotypic changes, which resemble those observed at the transition from the proB to small preB cell stage.

We next investigated whether the anti-Igβ mAb treatment could induce L chain gene rearrangements in μm-deficient mice. In order to examine the configuration of κL chain genes, CD45R (B220)[+] bone marrow cells were isolated from μm-deficient mice treated with either anti-Igβ mAb or PBS, and subjected to PCR analysis with a pair of primers specific for Vκ and Jκ2. Though low levels of PCR products corresponding to Vκ-Jκ1 and Vκ-Jκ2 joints could be detected even in control mice as described previously (Kitamura et al. 1992., Loffert et al. 1996), their levels were found much higher in the anti-Igβ-treated mice (Fig. 1). This result indicates that the cross-linking of Igβ on proB cells promotes the production of Vκ-Jκ joints in μm-deficient mice. To confirm the effect of Igβ cross-linking

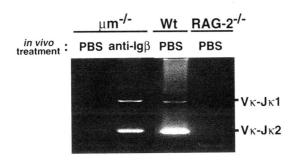

Fig. 1. The cross-linking of Igβ on μm-deficient proB cells in vivo activated L chain gene rearrangement. μm-deficient mice were injected intraperitoneally with either anti-Igβ mAb or PBS. On day 6 post-injection, bone marrow cells were isolated from the mice. B220⁺ cells were isolated from their bone marrow cells by means of magnetic cell sorting using B220-specific beads. Equivalent amounts of DNA extracted from B220⁺ cells were amplified by PCR with combination of Vκ degenerate primers and Jκ2 primer. As positive and negative controls, samples prepared from normal (Wt) and RAG-2-deficient (RAG-2⁻/⁻) mice in the same way were analyzed in parallel. Position of the amplified fragments corresponding to Vκ-Jκ1 and Vκ-Jκ2 joints are shown on the right side.

on L chain gene rearrangement, we took advantage of LMPCR assay that detects intermediates in the recombination reaction rather than its end-products (Schlissel et al. 1993). As shown in Fig. 2 (top), the frequency of intermediates of Vκ-Jκ rearrangement carrying broken-ended RSS upstream of Jκ1 greatly increased in the anti-Igβ treated μm-deficient mice as compared to the PBS-treated mice. Taken together, we concluded that Igβ signaling promotes V-J recombination reactions at κL chain loci.

The preBCR has also been suggested to generate a signal leading to inhibition of V_H to $D_H J_H$ recombination at the H chain locus, resulting in the allelic exclusion (Karasuyama et al. 1996). Therefore, LMPCR assay was further applied to examine recombination activity at the H chain locus by using primers specific for the broken-ended RSS upstream of $D_{FL16.1}$ (Schlissel et al. 1993; Constantinescu et al. 1997). The frequency of intermediates of V_H to $D_H J_H$ rearrangement was found much lower in CD45R(B220)⁺ bone marrow cells from the anti-Igβ treated mice than in those from PBS-treated control mice (Fig. 3, bottom). This clearly indicated that the cross-linking of Igβ on proB cells suppressed the V_H to $D_H J_H$ recombination at the H chain locus.

It was not yet clear whether the alteration of recombination activity at the H and L chain loci was a direct effect of the Igβ cross-linking or an indirect effect due to cellular differentiation. Therefore, we analyzed kinetics of the germline transcription at the H and L chain loci which has been demonstrated to correlate

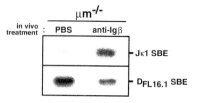

Fig. 2. The cross-linking of Igβ on proB cells in μm-deficient mice activated rearrangements at the L chain locus but suppressed those at the H locus. DNA extracted from B220⁺ bone marrow cells as in Fig. 1 was subjected to LMPCR to detect signal broken ends upstream of the Jκl segment (Jκl SBE) (upper panel) and those upstream of the D_{FL16.1} (DFL16.1 SBE) (lower panel) (Reproduced from the Journal of Experimental Medicine, 2000, 191, 1336. by copyright permission of the Rockefeller University Press).

strongly with rearrangements at those loci (Yancopoulos et al. 1985; Schlissel et al. 1989). Semi-quantitative RT-PCR revealed that the cross-linking of Igβ on proB cells induced drastic increase of κL chain germline transcripts which are derived from an unrearranged κL chain gene (Fig. 3A). In contrast, relative amounts of V_HJ558 germline transcripts derived from unrearranged V_H segments significantly diminished (Fig. 3B). We also found that the change of locus accessibility indicated by the up-regulation of κL chain germline transcripts and the down-regulation of V_H germline transcripts was detectable even prior to the cellular differentiation indicated by the change of surface phenotype (Maki et al. 2000). These results indicate that upon the anti-Igβ treatment, the κL chain locus in proB cells becomes open and more accessible for recombinases whereas the H chain locus becomes less accessible, even before the differentiation to the small preB cell stage.

Taken together, we concluded that the Igβ signaling induced a drastic change in the targeting of V(D)J recombinase activity, from being predominantly active at the H chain locus to being restricted to the L chain locus. Importantly, we also found that the change of locus accessibility at the H and L chain loci preceded the induction of cellular differentiation detected by the change of surface phenotype. Therefore, the alteration of locus accessibility induced by the Igβ signaling is not simply a consequence of the differentiation of proB cells to small preB cells in that decreased accessibility at the H chain locus and increased accessibility at the L chain locus have been reported (Schlissel et al. 1989; Schlissel et al. 1993; Constantinescu et al. 1997; Stanhope-Baker et al. 1996). Our results strongly suggest that the signaling through Igα/Igβ heterodimer of preBCR is involved in

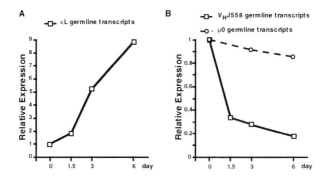

Fig. 3. The effect of the Igβ cross-linking on germline transcription from Ig genes. μm-deficient mice were injected intraperitoneally with 1 mg of anti-Igβ mAb. On day 0, 1.5, 3 and 6 post-injection, B220⁺ cells were enriched from their bone marrow as in Fig.1. DNase-treated RNA purified from the B220⁺ cells were reverse transcribed and amplified by PCR with primers specific to κL chain germline transcripts (A) and specific to V_HJ558 or μ0 germline transcripts (B). Southern blot of PCR products was hybridized with oligonucleotide probes specific to each gene. Radioactivity was quantified with Bio-image analyzer and normalized based on the expression of HRPT gene. The kinetics of expression levels of each transcript is shown. R elative radioactivity is plotted on the Y-axis. The signal in the B220⁺ cells on day 0 is set to 1.0 in each figure. (Reproduced from the Journal of Experimental Medicine, 2000, 191, 1337. by copyright permission of the Rockefeller University Press).

the regulation of the ordered Ig gene rearrangement by altering the accessibility of the H and L chain loci.

The system described in this study is unique in that the ordered program of Ig gene rearrangements can be triggered in vivo by one shot of the antibody injection. Thus, our system should provide a powerful tool for analyzing each process in ordered Ig gene rearrangements, such as the alteration of chromatin structure and germline transcription at a particular locus.

References

Alt FW, Rosenberg N, Lewis S, Thomas E, Baltimore D (1981) Organization and reorganization of immunoglobulin genes in A-MULV-transformed cells: rearrangement of heavy but not light chain genes. Cell 27:381-390

Constantinescu A, Schlissel MS (1997) Changes in locus-specific V(D)J recombinase activity induced by immunoglobulin gene products during B cell development. J Exp Med 185:609-620

Karasuyama H, Kudo A, Melchers F. (1990) The proteins encoded by the VpreB and λ5 pre-B cell-specific genes can associate with each other and with μ heavy chain. J Exp Med 172:969-972

Karasuyama H, Rolink A, Melchers F (1996) Surrogate light chain in B cell development. Adv Immunol 63:1-41

Karasuyama H, Rolink A, Shinkai Y, Young F, Alt FW, Melchers F (1994) The expression of Vpre-B/λ5 surrogate light chain in early bone marrow precursor B cells of normal and B cell-deficient mutant mice. Cell 77:133-143

Kitamura D, Rajewsky K (1992) Targeted disruption of μ chain membrane exon causes loss of heavy-chain allelic exclusion. Nature 356:154-156

Kitamura D, Roes J, Kuhn R, Rajewsky K (1991) A B cell-deficient mouse by targeted disruption of the membrane exon of the immunoglobulin μ chain gene. Nature 350:423-426

Koyama M, Ishihara K, Karasuyama H, Cordell JL, Iwamoto A, Nakamura T (1997) CD79α/CD79β heterodimers are expressed on pro-B cell surfaces without associated μ heavy chain. Int Immunol 9:1767-1772

Lassoued K, Nunez CA, Billips L, Kubagawa H, Monteiro RC, LeBlen TW, Cooper MD (1993) Expression of surrogate light chain receptors is restricted to a late stage in pre-B cell differentiation. Cell 73:73-86

Loffert D, Ehlich A, Muller W, Rajewsky K (1996) Surrogate light chain expression is required to establish immunoglobulin heavy chain allelic exclusion during early B cell development. Immunity 4:133-144

Maki K, Nagata N, Kitamura F, Takemori T, Karasuyama H (2000). Immunoglobulin β signaling regulates locus accessibility for ordered immunoglobulin gene rearrangements. J Exp Med 191: 1333-1340

Melchers F, Karasuyama H, Haasner D, Bauer S, Kudo A, Sakaguchi N, Jameson B, Rolink A (1993) The surrogate light chain in B-cell development. Immunol Today 14:60-68

Nagata K, Nakamura T, Kitamura F, Kuramochi S, Taki S, Campbell KS, Karasuyama H (1997) The Igα/Igβ heterodimer on μ-negative proB cells is competent for transducing signals to induce early B cell differentiation. Immunity 7:559-570

Rajewsky K (1996) Clonal selection and learning in the antibody system. Nature 381:751-758

Reth MG, Ammirati P, Jackson S, Alt FW (1985) Regulated progression of a cultured pre-B-cell line to the B-cell stage. Nature 317:353-355

Schlissel MS, Baltimore D (1989) Activation of immunoglobulin kappa gene rearrangement correlates with induction of germline kappa gene transcription. Cell 58:1001-1007

Schlissel M, Constantinescu A, Morrow T, Baxter M, Peng A (1993) Double-strand signal sequence breaks in V(D)J recombination are blunt, 5'-phosphorylated, RAG-dependent, and cell cycle regulated. Genes Dev 7:2520-2532

Shinkai Y, Rathbun G, Lam KP, EOltz EM, Stewart V, Mendelsohn M, Charron J, Datta M, Young F, Stall AM, Alt FW (1992) RAG-2-deficient mice lack mature lymphocytes owing to inability to initiate V(D)J rearrangement. Cell 68:855-867

Stanhope-Baker P, Hudson M, Shaffer AL, Constantinescu A, Schlissel MS (1996) Cell type-specific chromatin structure determines the targeting of V(D)J recombinase activity in vitro. Cell 85:887-897

Tsubata T, Reth M (1990) The products of pre-B cell-specific genes (λ5 and VpreB) and the immunoglobulin μ chain form a complex that is transported onto the cell surface. J Exp Med 172:973-976

Yancopoulos GD, Alt FW (1985) Developmentally controlled and tissue-specific expression of unrearranged VH gene segments. Cell 40:271-281

20
Regulation of phospolipase C-γ2 and phosphoinositide 3-kinase by adaptor proteins in B cells

Tomohiro Kurosaki

Department of Molecular Genetics, Institute for Liver Research, Kansai Medical University, Moriguchi 570-8506, Japan

Summary. The importance of phosphoinositide 3-kinase (PI3K) and phospholipase C (PLC)-γ2 in B cell development and B cell receptor (BCR) signaling has been highlighted by gene targeting experiments in mice. Based on these evidence, the molecular connections between BCR and these effector enzymes have been intensively studied. This mechanism involves the action of cytoplasmic adaptor molecules such as BLNK and BCAP, which participate in forming multicomponent signaling complexes and thereby directing the appropriate subcellular localization of effector enzymes, PI3K and PLC-γ2.

Key words. BCR, BCAP, BLNK, CD19, PI3-K, PLC-γ2

Introduction

The BCR is a multiprotein complex consisting of the membrane-bound immunoglobulin (mIg) molecule and the Igα/Igβ heterodimer (CD79α, β). Antigen is bound by the variable domains of the mIg heavy and light chains, whereas coupling of the receptor to intracellular signaling proteins is achieved by the Igα/Igβ heterodimer carrying immunoreceptor tyrosine-based activation motifs (ITAMs) in their cytoplasmic domains. An early step in BCR signal transduction is the activation of protein tyrosine kinases (PTKs) that phosphorylate several substrate proteins including the Igα/Igβ ITAMs. To date, three distinct families of cytoplasmic PTKs, Src-family PTK (including Lyn, Fyn, and Blk), Syk, and Btk, have been identified that are required for BCR activation and B cell development. These PTKs are sequentially activated to regulate the coordinated generation of second messengers, including IP_3 (inositol [1,4,5]trisphosphate) and $PI(3,4,5)P_3$ (phosphatidylinositol[3,4,5]trisphosphate).

The IP$_3$ generation requires the enzymatic activation of phospholipase C (PLC)-γ2, while phosphoinositide 3-kinase (PI3K) is responsible for the production of PI(3,4,5)P$_3$ (Kurosaki 1999).

Coupling mechanisms between PTKs and effectors

Based upon the evidence that BCR-associated PTKs lie upstream of PI3K and PLC-γ2 in BCR signaling, the coupling mechanisms between PTKs and PI3K/PLC-γ2 have been extensively studied. Data emerging from several laboratories demonstrate that adaptor molecules play key roles in coupling the BCR-activated PTKs to PI3K/PLC-γ2 activation.

BLNK was isolated as one of rapidly tyrosine phosphorylated proteins upon BCR cross-linking (Fu et al. 1998), and turned out to be a critical adaptor for PLC-γ2 activation (Ishiai et al. 1999a). BCR-induced BLNK phosphorylation is apparently mediated by Syk, since this phosphorylation is lost in Syk-deficient DT40 B cells (Fu et al. 1998), implicating the scenario that after being phosphorylated by Syk, BLNK provides the docking site for PLC-γ2 SH2 domains, which in turn is critical for translocation of PLC-γ2 to the plasma membrane and its subsequent activation. Consistent with this model, PLC-γ2 SH2 mutant fails to associate with the phosphorylated BLNK as well as to be translocated to the plasma membrane upon BCR cross-linking (Ishiai et al. 1999b). In addition to Syk, Btk was shown to be required for PLC-γ2 activation. The idea that Btk, like Syk, participates in BLNK phosphorylation was first thought to account for the requirement for Btk. However, this was shown to be unlikely. Instead, recent identification of BLNK as one of the major Btk SH2-binding proteins (Hashimoto et al. 1999; Su et al. 1999) has evoked the idea that BLNK, phosphorylated by Syk, provides docking sites for Btk as well as PLC-γ2,

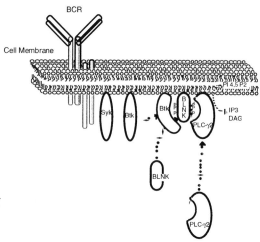

Fig. 1. Model for the mechanism of BLNK-mediated PLC-γ2 activation upon BCR stimulation.

thereby bringing Btk into close proximity with PLC-γ2. Then, the activated Btk phosphorylates PLC-γ2, leading to its activation (Fig. 1).

In addition to PLC-γ2, BCR cross-linking activates heterodimer-type (class Ia) PI3K, composed of a catalytic subunit (p110) and a regulatory subunit (p85α, p85β, or p55γ) (Fruman et al. 1998). Among regulatory subunits, p85α is predominantly expressed in B cells. Two SH2 domains of the p85α are likely to be important for PI3K activation mechanism. Both SH2 domains bind with high affinity to phosphotyrosines in the YxxM motif and this SH2-dependent binding to docking proteins is thought to be critical for recruitment of PI3K to the plasmam membrane and its subsequent activation (Buhl and Cambier 1999; Tuveson et al. 1993). Since the transmembrane protein CD19 has two YxxM motifs in its cytoplamic domain, CD19 has been thought to be involved in docking between BCR and p85α. However, because B cell defects in p85α[-/-] mice are more severe than those in CD19[-/-] mice (Fruman et al. 1999; Suzuki et al. 1999), this CD19-mediated mechanism seems not to fully account for the activation mode of the PI3K. Based upon the hypothesis that additional adaptor molecules could participate in linking between BCR-associated PTKs and PI3K, we have recently cloned a novel cytoplasmic adaptor molecule, named BCAP, containing four YxxM motifs. Tyrosine phosphorylation of BCAP is mediated by Syk, thereby providing binding site(s) for p85α SH2 domains. The functional importance of BCAP in PI3K activation has been underscored by disruption of the BCAP gene in DT40 B cells; BCR-mediated PIP_3 generation is inhibited in the deficient cells (Okada et al. 2000). Collectively, BCAP plays an important role in coupling between Syk and PI3K in BCR signaling. However, even in the absence of BCAP, the residual PIP_3 generation occurs, indicating the involvement of additional adaptor molecules such as CD19 and Gab in full PI3K activation (Fig. 2).

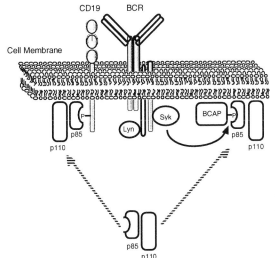

Fig. 2. Model for the mechanism of PI3K activation upon BCR stimulation. PI3K is composed of a catalytic subunit (p110) and a regulatory subunit (p85).

In vivo functions of PI3K and PLC-γ2

Deficiencies of p85 (Fruman et al. 1999; Suzuki et al. 1999) and PLC-γ2 (Hashimoto et al. 2000; Wang et al. 2000) result in a greater reduction of B1 B cells, while the number of mature conventional B (B2) cells is moderately decreased in these knock-out mice. B1 cells are predominantly located in the peritoneal cavity, where they can be distinguished from the B2 cells as IgM^{hi}IgD^{lo}CD23$^-$CD43$^+$ and CD5$^+$ (B1-a) or CD5$^-$ (B1-b) cells. Although it still remains unresolved whether B2 B cells require ligand-mediated positive selection for their maintenance in the periphery, the compelling evidence for the role of self-antigens in positive selection in B1 cells has come from a recent study of Hayakawa et al (Hayakawa et al. 1999). Hence, the simple explanation for the reduction of B1 cells in PLC-γ2$^{-/-}$ and p85α$^{-/-}$ mice is that both enzymes are required for this self-ligand-induced BCR signaling, thereby contributing to induction and/or maintenance of B1 cells.

PLC-γ2$^{-/-}$ and p85α$^{-/-}$ mice exhibit poor antibody responses to T-independent type II (TI-2) antigens such as NP-Ficoll, but normal responses to T-dependent (TD) antigens (Fruman et al. 1999; Hashimoto et al. 2000; Suzuki et al. 1999; Wang et al. 2000). In addition to the involvement of B1 cells in TI-2 responses, a recent study using Pky-2$^{-/-}$ mice suggests an important role for marginal zone B cells in these responses (Guinamard et al. 2000). These mice lack splenic marginal zone B cells and exhibit a deficit in IgM and IgG3 responses to the NP-Ficoll. Marginal zone B cells have a distinct character compared to the more common follicular B cells, including higher expression of complement receptor (CD21^{hi}) and the nonclassical MHC molecule CD1^{hi}. The splenic marginal zone B cells develop normally in Btk-deficient mice. However, some clones of antigen-specific marginal zone B cells require Btk for their development or survival, implicating that they may be positively selected, like B-1 cells, via the BCR (Martin and Kearney 2000). Because Btk, PLC-γ2, and PI3K are thought to function in a common BCR signaling pathway, requirement for PLC-γ2/PI3K in development and/or activation of positively selected marginal zone B cells may account for their requirement in TI-2 responses.

References

Buhl AM, Cambier JC (1999) Phosphorylation of CD19 Y484 and Y515, and linked activation of phosphatidylinositol 3-kinase, are required for B cell antigen receptor-mediated activation of Bruton's tyrosine kinase. J Immunol 162:4438-4446

Fruman DA, Meyers RE, Cantley LC (1998) Phosphoinositide kinases. Annu Rev Biochem 67:481-507

Fruman DA, Snapper SB, Yballe CM, Davidson L, Yu JY, Alt FW, Cantley LC (1999) Impaired B cell development and proliferation in absence of phosphoinositide 3-kinase p85α. Science 283:393-397

Fu C, Turck CW, Kurosaki T, Chan AC (1998) BLNK: a central linker protein in B cell activation. Immunity 9:93-103

Guinamard R, Okigaki M, Schlessinger J, Ravetch VJ (2000) Absence of marginal zone B cells in Pyk-2-deficient mice defines their role in the humoral response. Nature Immunol 1:31-36

Hashimoto A, Takeda K, Inaba M, Sekimata M, Kaisho T, Ikehara S, Homma Y, Akira S, Kurosaki T (2000) Essential role of phospholipase C-γ2 in B cell development and function. J Immunol 165:1738-1742

Hashimoto S, Iwamatsu A, Ishiai M, Okawa K, Yamadori T, Matsushita M, Baba Y, Kishimoto T, Kurosaki T, Tsukada S (1999) Identification of the SH2 domain binding protein of Bruton's tyrosine kinase as BLNK-functional significance of Btk-SH2 domai in B-cell antigen receptor-coupled calcium signaling. Blood 94:2357-2364

Hayakawa K, Asano M, Shinton AS, Gui M, Allman D, Stewart LC, Silver J, Hardy R R (1999) Positive selection of natural autoreactive B cells. Science 285:113-116

Ishiai M, Kurosaki M, Pappu R, Okawa K, Ronko I, Fu C, Shibata M, Iwamatsu A, Chan AC, Kurosaki T (1999a) BLNK required for coupling Syk to PLCγ2 and Rac1-JNK in B cells. Immunity 10:117-125

Ishiai M, Sugawara H, Kurosaki M, Kurosaki T (1999b) Association of phospholipase C-γ2 Src homology 2 domains with BLNK is critical for B cell antigen receptor signaling. J Immunol 163:1746-1749

Kurosaki T (1999) Genetic analysis of B cell antigen receptor signaling. Annu Rev Immunol 17:555-592

Martin F, Kearney FJ (2000) Positive selection from newly formed to marginal zone B cells depends on the rate of clonal production, DC19, and *btk*. Immunity 12:39-49

Okada T, Maeda A, Iwamatsu A, Gotoh K, Kurosaki T (2000) BCAP:The tyrosine kinase substrate that connects B cell receptor to phosphoinositide 3-kinase activation. Immunity 13:817-827

Su YW, Zhang Y, Schweikert J, Koretzky GA, Reth M, Wienands J (1999) Interaction of SLP adaptors with the SH2 domain of Tec family kinases. Eur J Immunol 29:3702-3711

Suzuki H, Terauchi Y, Fujiwara M, Aizawa S, Yazaki Y, Kadowaki T, Koyasu S (1999) *Xid*-like immunodeficiency in mice with disruption of the p85α subunit of phosphoinositide 3-kinase. Science 283: 390-392

Tuveson DA, Carter RH, Soltoff SP, Fearon DT (1993) CD19 of B cells as a surrogate kinase insert region to bind phosphatidylinositol 3-kinase. Science 260: 986-989

Wang D, Feng J, Wen R, Marine J-C, Sangster YM, Parganas E, Hoffmeyer A, Jackson WC, Cleveland LJ, Murray JP, Ihle NJ (2000) Phospholipase Cγ2 is essential in the functions of B cell and several Fc receptors. Immunity 13:25-35

21
Fc receptor signaling during phagocytosis

Erick García-García and Carlos Rosales[1]

Immunology Department, Instituto de Investigaciones Biomédicas, Universidad Nacional Autónoma de México, Apto. Postal 70228, Cd. Universitaria, México D.F. - 04510, Mexico

Summary. The molecular machinery involved in Fc receptor-mediated phagocytosis in the different cell types of the immune system is still poorly defined. Cross-linking of FcγR results in activation of Src family kinases followed by activation of Syk family kinases. After Syk activation several phosphorylated proteins have been identified including: phospholipase C (PLC), phosphatidylinositol 3-kinase (PI 3-K), extracellular signal-regulated kinase (ERK), and GTPases of the Rho family. For phagocytosis, the involvement of PI 3-K and ERK seems to be dependent on the cell type. FcγR-mediated phagocytosis by monocytic cells (THP-1) was not blocked by wortmannin, a specific inhibitor of PI 3-K, nor by PD98059, a specific inhibitor of ERK. However, upon differentiation of THP-1 monocytes to a macrophage phenotype both wortmannin and PD98059 efficiently blocked FcγR-mediated phagocytosis. Additionally, phagocytosis by neutrophils, a more efficient phagocyte, was inhibited both by wortmannin and PD98059. Neutrophils and macrophage-differentiated monocytes presented significantly more efficient phagocytosis than monocytes, upon PMA stimulation. Taken together, these results indicate that less efficient phagocytic leukocytes, such as monocytes, do not require PI 3-K and ERK for phagocytosis. However, upon differentiation into macrophages ERK and PI 3-K are recruited as part of the phagocytic machinery.

Key words. Monocyte, Macrophage, Fc receptors, Phagocytosis, Signal transduction

---

[1] Supported by grants: IN201797 DGAPA-UNAM, and 31088-M CONACYT, Mexico.

Introduction

The various mechanisms used by multicellular organisms to protect themselves from infection by viruses, bacteria, fungi, and protozoa are collectively known as host defense. Among the defense mechanisms taking place during the inflammatory process there are early systems, such as complement, that are very rapid, yet poorly specific. Phagocytic leukocytes represent a more advanced defense mechanism to detect and eliminate pathogens. Finally, the specific recognition of antigens by T and B lymphocytes is the most advanced and precise mechanism of host defense. However, none of these systems is by itself completely effective in protecting the host. Instead, all these systems work in cooperation. Antibodies produced by the adaptive specific immune system help to activate complement and to trigger phagocytosis. Because phagocytosis is the mechanism by which most of the potential pathogens are ultimately destroyed (Jones et al. 1999) there is a lot of interest in elucidating the biochemical signals that regulate this important function of host defense (Greenberg 1999). This chapter will review the signal transduction pathways leading to phagocytosis that are activated by antibodies on the various phagocytic leukocytes.

Fc Receptors

Antibodies present two main functions in host defense: the binding to antigen via their antigen-combining sites, and the mobilization of cellular defense mechanisms via their carboxyl terminal Fc portion. Antibodies are thus a bridge between the specific recognition of immune system, and the not so specific, but highly destructive mechanisms of phagocytic leukocytes. Antibodies are recognized by these cells through specific receptors for their Fc portion. Receptors for the Fc portion of immunoglobulin G molecules, known as Fc gamma receptors (FcγR), can trigger various functions in many cells of the immune system. These include phagocytosis, antibody dependent cell-mediated cytotoxicity, generation of the respiratory burst, and production of inflammatory mediators and cytokines (Daeron 1997).

Three classes of FcγR have been identified, FcγRI (CD64), FcγRII (CD32), and FcγRIII (CD16) (Fig. 1). They are coded for by different genes and differ in their relative avidity for IgG, molecular structure, and cellular distribution (Ravetch 1997). FcγRI binds monomeric IgG while FcγR types II and III present only avidity for multimeric immune complexes. FcγRI is expressed on monocytes, macrophages, and interferon-γ (IFN-γ)-stimulated neutrophils. FcγRI is associated with a dimer of gamma chains. Each γ chain contains tyrosine residues that become phosphorylated upon receptor activation. These tyrosine residues are found within a common motif identified in many chains of antigen receptors (such as TCR and BCR), and Fc receptors. This motif is known as ITAM, for immunoreceptor tyrosine-based activation motif. There are several isoforms of FcγRII, derived from its three genes and from alternative splicing. FcγRII isoforms have different distribution in hematopoietic cells. FcγRIIA is found mainly in phagocytic cells (neutrophils, monocytes, and macrophages), whereas FcγRIIB is expressed in B and T lymphocytes. FcγRII does not have associated γ chains. FcγRIIA contains an ITAM in its cytoplasmic portion, while FcγRIIB has a different tyrosine-containing motif involved in negative signaling. This motif is known as ITIM, for immunoreceptor tyrosine-based inhibition motif (see chapter by M. Daeron, this volume). FcγRIII has two isoforms: FcγRIIIA is a receptor with a transmembrane portion and a cytoplasmic tail, associated with a heterodimer of γ/ζ chains, containing ITAMs. It is expressed mainly of natural killer (NK) cells and macrophages. FcγRIIIB is present exclusively on neutrophils and it is a glycosylphosphatidylinositol (GPI)-linked receptor, lacking a cytoplasmic tail. No other subunits are known to associate with it, but it may signal in cooperation with other receptors (Daeron 1997) (Fig. 1).

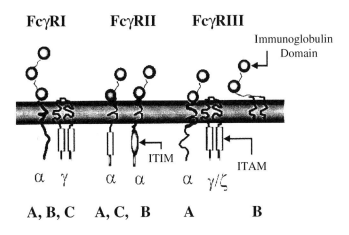

Fig. 1. Fcγ receptor family. The family of receptors for the Fc portion of immunoglobulin G molecules comprises three members FcγRI, FcγRII, and FcγRIII. Receptors are coded for by several genes designated with capital letters (A, B, or C). Each receptor subunit is designated with Greek letters (α, γ, or ζ). ITAM, immunoreceptor tyrosine-based activation motif. ITIM, immunoreceptor tyrosine-based inhibition motif.

Signal transduction by FcγR

Activation of FcγR, as well as other immunoreceptors (such as TCR, BCR, and FcεRI), results in common molecular events involving activation of Src family kinases followed by activation of Syk family kinases (van Leeuwen and Samelson 1999). In phagocytes the kinases Src, Fyn, Fgr, Hck, and Lyn have all been identified. The particular kinases involved in FcγR signaling depend on the particular ITAM present on the cytoplasmic portion of each receptor (see the chapter by J. Ravetch, this volume). Due to their lipid anchor to the cell membrane these kinases are in close proximity to the cytoplasmic tails of FcγRs, though they remain inactive until FcγRs are cross-linked. Once activated, the SH2 domain of the kinase is free and can bind to the phosphotyrosine residues in the ITAM of the engaged FcγR (Fig. 2). Another kinase, Syk, is involved in FcγR signaling. Syk belongs to the ZAP-70 kinase family. These kinases are not myristoylated and therefore are exclusively cytoplasmic. Syk is present in all hematopoietic cells, whereas ZAP-70 is expressed in T cells and NK cells (Chu et al. 1998). Syk also binds, via their SH2 domains, to the phosphorylated ITAMs of FcγRs. There, the kinase activity of Syk is stimulated. Syk kinase is now recognized as a central element in signaling from FcγR to initiate several responses, including phagocytosis.

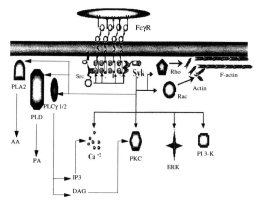

Fig. 2. Fcγ receptor signal transduction. FcγR cross-linking induces activation of Src family kinases and Syk family kinases. These enzymes associate with phosphorylated ITAMs in the cytoplasmic tails of the receptor. Syk activation leads to phosphorylation and activation of several substrates involved in regulation of phagocytosis and other responses. Among them are phospholipase C (PLC) γ1 and γ2 which produce inositoltrisphosphate (IP3) and diacylglycerol (DAG). These second messengers cause calcium release and activation of protein kinase C (PKC), respectively. Phospholipase A2 (PLA2), which produces arachidonic acid (AA), a precursor of some inflammatory mediators; phospholipase D (PLD), which produces phosphatidic acid (PA); phosphatidylinositol 3-kinase (PI 3-K); extracellular signal-regulated kinase (ERK); and GTPases of the Rho and Rac family, which are involved in the regulation of the actin cytoskeleton.

The downstream events of Syk activation are not clearly defined at the moment. However, a series of phosphorylated proteins and activated enzymes have been detected after FcγR stimulation. Among these activated enzymes are: phospholipase C (PLC) γ1 and γ2, which are responsible for induction of the second messengers inositol 1,4,5-trisphosphate (IP$_3$) and diacylglycerol (DAG); phosphatidylinositol 3-kinase (PI 3-K); paxillin, a cytoskeletal protein; extracellular signal-regulated kinase (ERK), also known as mitogen-activated protein kinase (MAPK); and GTPases of the Rho and Rac family, which are involved in regulation of the actin cytoskeleton (Daeron 1997; Sánchez-Mejorada and Rosales 1998) (Fig. 2). The particular role that each of these molecules plays in FcγR signal transduction is currently the focus of active research.

Phagocytosis

Phagocytosis is the process of recognition and engulfment of microorganisms or tissue debris that accumulate at sites of infection and inflammation. This function, essential for successful host defense, is performed most efficiently by migrating leukocytes denominated "professional phagocytes". These include neutrophils, monocytes, and macrophages. Neutrophils (PMN) are the most abundant leukocytes in blood, and are the first cells to arrive at sites of inflammation. Monocytes are the circulating precursors of macrophages. During inflammation monocytes migrate through the endothelium of blood vessels into the tissues. There, under the influence of cytokines such as IFN-γ or colony stimulating factors (CSF), monocytes differentiate into macrophages. Monocytes respond to various stimuli by producing cytokines, thus regulating the inflammatory process. Although, monocytes are phagocytic leukocytes, they are not as efficient phagocytes as macrophages. This difference in cell behavior suggests that differentiation of monocytes into macrophages involves at the same time, changes in their biochemical signaling pathways, so as to deliver different responses.

Phagocytosis starts when an invading microorganism is detected by specific receptors on the cell membrane. The three classes of FcγR allow phagocytes to recognize IgG-coated microorganisms and other IgG-coated particles (Jones et al. 1999).

Various experimental approaches have been used to explore the phagocytic potential of the different FcγRs. These include transfection of FcγRs in several cell types, selective stimulation of individual FcγR with specific monoclonal antibodies, and genetic elimination of some FcγRs. Upon transfection into COS-1 cells, FcγRI did not render these cells phagocytic, though they bound IgG-coated erythrocytes efficiently. However, co-expression of the associated γ-chain rendered FcγRI functional for phagocytosis. FcγRIIA easily mediated phagocytosis in transfected cells, whereas its isoform FcγRIIB did not. Tyrosine kinase inhibitors and mutations in the tyrosine residues of the ITAMs of FcγRIIA drastically reduced the phagocytic activity of this receptor. The case of FcγRIII is interesting because its two isoforms are very different. FcγRIIIA transfected into COS-1 cells was capable of sending a phagocytic signal in the absence of other FcγRs, but the presence of the γ-chain was required both for membrane expression and function. FcγRIIIB, despite its absence of transmembrane and cytoplasmic regions, is capable of initiating signal transduction events such as calcium release and actin polymerization. The way FcγRIIIB transduces a signal is not clear, but it is thought that this receptor associates with other molecules on the cell membrane to initiate phagocytosis. FcγRIIA has been suggested to be one of these molecules, and the complement receptor type 3 (CR3, Mac-1) to be another. The mechanisms by which FcγRIIIB may recruit the signaling capabilities of FcγRIIA, Mac-1, or both, remain unknown.

Data described above clearly indicate that representative members from each FcγR class are capable of triggering IgG-mediated phagocytosis and, although they have particular requirements, in all cases (except FcγRIIIB) phosphorylation of ITAM sequences in the cytoplasmic tail of their α-chains, or their associated γ-subunits is a constant requirement (Sánchez-Mejorada and Rosales 1998).

Role of ERK and PI 3-K in FcγR-mediated phagocytosis

One of the major cellular responses initiated by FcγR cross-linking, specially in myelomonocytic cells and in neutrophils (PMN), is phagocytosis (Jones et al. 1999). The molecular machinery needed for this function is a matter of great interest and active research (Greenberg 1999). Several enzymes have been suggested to participate in various FcγR-mediated responses. In particular, ERK is required for several FcγR-mediated functions in various cell types. However, participation of this molecule in the phagocytic process is not clearly defined. There are reports indicating that ERK is needed for phagocytosis of IgG-opsonized particles by PMN. But there are also reports indicating that ERK is not required for phagocytosis by monocytes. So, it seems that ERK may be involved in phagocytosis in some cases but not in others, depending on the cell type.

Similarly, PI 3-K has been reported to be an important molecule during FcγR-mediated phagocytosis. The role of PI 3-K in this function seems to be modulation of the assembly of the submembranous actin filament system and membrane redistribution leading to particle internalization (Lennartz 1999). Reports showing a role for PI 3-K in phagocytosis used professional phagocytic cells, such as PMN and macrophages. For less efficient phagocytes, such as monocytes, there are not reports regarding the involvement of PI 3-K during phagocytosis. These reports also suggested that PMN and macrophages present more efficient phagocytosis because they use PI 3-K and ERK for this function, while less efficient phagocytes, like monocytes do not.

To test the hypothesis that differentiation of monocytes into a more macrophage phenotype, is accompanied by recruitment of ERK and possibly PI 3-K to the phagocytic machinery, we assessed the participation of these signaling molecules in phagocytosis of IgG-coated erythrocytes (EIgG) by monocytic cells and by macrophage-differentiated monocytes. Phagocytosis of EIgG by PMN was dependent on PI 3-K and ERK, since the PI 3-K inhibitor (wortmannin) and also the ERK inhibitor (PD98059) completely blocked it (Fig. 3). In contrast, neither wortmannin nor PD98059 had any effect on EIgG phagocytosis by monocytes (Fig. 3).

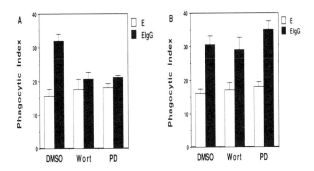

Fig. 3. PI 3-K and ERK are not required for monocyte FcγR-mediated phagocytosis. Neutrophils (A) or THP-1 monocytic cells (B) were mixed with sheep erythrocytes (E) or IgG-opsonized sheep erythrocytes (EIgG), and incubated at 37 °C to allow ingestion of erythrocyte targets. Cells were treated with 50 nM wortmannin (Wort), 30 μM PD98059 (PD), or only the solvent (DMSO) before mixing with the erythrocyte targets. Data are reported as phagocytic index (erythrocytes ingested by 100 phagocytes). Data are mean ± S.E. of 12 independent determinations.

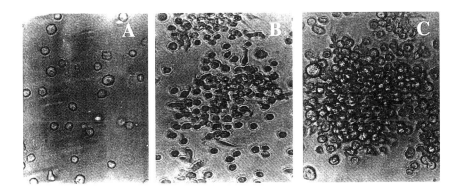

Fig. 4. Retinoic acid and IFN-γ induce differentiation of THP-1 monocytic cells into macrophages. THP-1 cells were cultured in regular RPMI-1640 medium (A), or in medium supplemented with 1 μM retinoic acid for 48 hours, and then with 15 ng/ml IFN-γ for additional 24 hours (B), or 48 hours (C).

These results indicated that efficient phagocytes (PMN in this case) use PI 3-K and ERK for phagocytosis, while monocytes do not. To test the idea that differentiation of monocytes into a more macrophage phenotype, is accompanied by recruitment of PI 3-K and ERK to the phagocytic machinery, the participation of these signaling molecules in phagocytosis of EIgG by macrophage-differentiated monocytes was assessed. THP-1 monocytic cells were differentiated into macrophages by culturing them with a combination of retinoic acid and IFN-γ. This treatment results in macrophage-differentiated THP-1 cells that grow, contrary to undifferentiated cells, attached to the tissue culture flask (Fig. 4).

Twenty four-hour differentiated THP-1 cells were collected and used in FcγR-mediated phagocytosis assays. Treatment with wortmannin before allowing EIgG ingestion, did not block phagocytosis. However, treatment with PD98059 effectively inhibited phagocytosis (Fig. 5A), thus indicating that by this time the phagocytic machinery was already dependent on ERK. By 48 hours of differentiation, both wortmannin and PD98059 completely blocked EIgG phagocytosis (Fig. 5B). These results clearly indicated that now both PI 3-K and ERK were required for phagocytosis and suggested that complete differentiation of monocytic cells into a macrophage phenotype results in recruitment of both PI 3-K and ERK into the phagocytic machinery.

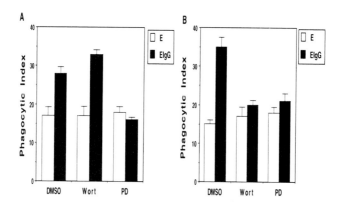

Fig. 5. PI 3-K and ERK are recruited for FcγR-mediated phagocytosis upon monocyte to macrophage differentiation. THP-1 monocytic cells were cultured with 1 μM retinoic acid for 48 h, and then with 15 ng/ml IFN-γ for additional A) 24 hours, or B) 48 hours. Cells were then trypsinized, washed and mixed with sheep erythrocytes (E) or IgG-opsonized sheep erythrocytes (EIgG), and then incubated at 37°C to allow ingestion of erythrocyte targets. Cells were treated with 50 nM wortmannin (Wort), 30 μM PD98059 (PD), or only the solvent (DMSO) before mixing with erythrocytes. Data are reported as phagocytic index (erythrocytes ingested by 100 phagocytes). Data are mean ± S.E. of eight independent determinations.

The results presented above demonstrate that efficient phagocytes (PMN and macrophages) use PI 3-K and ERK for FcγR-mediated phagocytosis, whereas monocytes do not. It is also known that professional phagocytes present low levels of basal phagocytosis. However, during the inflammatory response they become activated by different stimuli, and then present a more efficient phagocytic response. The results presented above also suggested that the use of PI 3-K and ERK for phagocytosis would render PMN and macrophage-differentiated monocytes more efficient for this process. To confirm this idea phagocytes were stimulated with 100 ng/ml phorbol myristate acetate (PMA) during phagocytosis assays. PMN and macrophage-differentiated monocytes presented significantly higher phagocytic activity than monocytes (Fig. 6). Phagocytic indexes for PMN, monocytes, and macrophage-differentiated monocytes were five, two, and four fold higher than the respective indexes of unstimulated control cells. Taken together these results support the idea that efficient phagocytes require ERK and PI 3-K for FcγR-mediated phagocytosis.

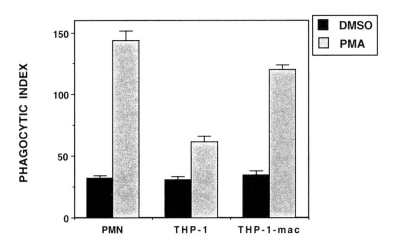

Fig. 6. Upon stimulation PMN and macrophage-differentiated monocytes show a stronger phagocytic response. Neutrophils (PMN), monocytic cells (THP-1) or macrophage-differentiated monocytes (THP-1-mac) were treated with 100 ng/ml phorbol myristate acetate (PMA) or only the solvent (DMSO), and then mixed with IgG-opsonized sheep erythrocytes (EIgG). Phagocytosis of erythrocyte targets is reported as phagocytic index (erythrocytes ingested by 100 phagocytes). Data are mean ± S.E. of four independent determinations.

Conclusion

Cross-linking of FcγR on the cell membrane by multivalent antigens triggers intracellular signaling, which initiates with the activation of the Src and Syk families of tyrosine kinases (Fig. 2). Syk seems to be the central kinase for most FcγR-mediated responses. After Syk activation a series of different substrates have been identified in the various cell types. For phagocytosis, the enzymes ERK and PI 3-K seem to be required for ingestion by more efficient phagocytes, such as PMN and macrophages, but not by monocytes. Upon differentiation to a macrophage phenotype, phagocytosis by THP-1 cells became sensitive to wortmannin, a specific inhibitor of PI 3-K, and to PD98059, a specific inhibitor of ERK. After stimulation with PMA both neutrophils and macrophage-differentiated monocytes presented significantly higher phagocytosis than monocytes. These data support the idea that professional phagocytes require both ERK and PI 3-K to accomplish their phagocytic functions, and that these enzymes are added to the phagocytic machinery upon monocyte differentiation.

References

Chu DH, Morita CT, Weiss A (1998) The Syk family of protein tyrosine kinases in T-cell activation and development. Immunol Rev 165: 167-180

Daeron M (1997) Fc receptor biology. Annu Rev Immunol 15: 203-234

Greenberg S (1999) Modular components of phagocytosis. J Leukoc Biol 66: 712-717

Jones SL, Lindberg FP, Brown EJ (1999) Phagocytosis. In: Paul WE (eds) Fundamental Immunology. Lippincott-Raven Publishers Philadelphia, pp 997-1020

Lennartz MR (1999) Phospholipases and phagocytosis: the role of phospholipid-derived second messengers in phagocytosis. Int J Biochem Cell Biol 31: 415-430

Ravetch JV (1997) Fc receptors. Curr Opin Immunol 9: 121-125

Sánchez-Mejorada G, Rosales C (1998) Signal transduction by immunoglobulin Fc receptors. J Leukoc Biol 63: 521-533

van Leeuwen JEM, Samelson LE (1999) T cell antigen-receptor signal transduction. Curr Opin Immunol 11: 242-248

22
Molecular mechanism of paired immunoglobulin-like receptor B (PIR-B)-mediated inhibitory signal

Akito Maeda[1#], Andrew M. Scharenberg[2], Satoshi Tsukada[3], Joseph B. Bolen[4], Jean-Pierre Kinet[2], and Tomohiro Kurosaki[1]

[1]Department of Molecular Genetics, Institute for Liver Research, Kansai Medical University, Moriguchi 570-8506, Japan
[2]Laboratory of Allergy and Immunology, Beth Israel Hospital and Harvard Medical School, Boston, Massachusetts, 02215, USA
[3]Department of Medicine III, Osaka University Medical School, Osaka 565, Japan
[4]Hoechst Marion Roussel, Bridgewater, New Jersey, 08807, USA
[#]Current address: Department of Molecular Immunology and Allergology, Kyoto University Medical School, Kyoto, 606-8501, Japan

Summary. B cell express paired immunoglobulin-like receptor B (PIR-B) contains four potential ITIMs (immunoreceptor tyrosine-based inhibitory motifs) in the cytoplasmic domain. Coligation of PIR-B with B cell antigen receptor (BCR) blocks antigen-induced B cell activation. This inhibition is mediated in part by recruitment of SH2-containing tyrosine phosphatases SHP-1 and SHP-2 to the phosphorylated ITIMs in the cytoplasmic domain of PIR-B. PIR-B ligation inhibits the BCR-induced tyrosine phosphorylation of Igα/Igβ, Syk, Btk, and phospholipase C (PLC)-γ2. Overexpression of a catalytically inactive form of SHP-1 prevents the PIR-B-mediated inhibition of tyrosine phosphorylation of Syk, Btk, and PLC-γ2. Dephosphorylation of Syk and Btk mediated by SHP-1 leads to a decrease of their kinase activity, which in turn inhibits tyrosine phosphorylation of PLC-γ2. In addition, Lyn is required for tyrosine phosphorylation of PIR-B.

Key words. PIR-B, ITIM, SHP-1, SHP-2

Introduction

B cells express PIR-B, an integral membrane protein, that contains six Ig-like domains in the extracellular portion and four potential ITIMs in the cytoplasmic domain (Hayami et al.1997; Kubagawa et al.1997). We and others have recently demonstrated that PIR-B inhibits the BCR signaling by a phosphotyrosine-dependent manner (Blery et al. 1998; Maeda et al. 1998; Maeda et al. 1999). Although the ligand for PIR-B is still unknown, PIR-B is inert until co-ligated to BCR at which time it undergoes tyrosine phosphorylation of ITIMs in its cytoplasmic domain. This phosphorylation leads to the recruitment of SHP-1 and SHP-2 capable of inhibiting calcium mobilization in the stimulated B cells (Maeda et al. 1998).

Tyr^{771} and Tyr^{810} are required for PIR-B-mediated inhibition of calcium mobilization and for recruitment of SHP-1 and SHP-2

Chimeric receptor FcγRIIB-PIR-B with the cytoplasmic domain of PIR-B and the extracellular domain of mouse FcγRIIB inhibits activation of B cells and mast cells like native PIR-B(Blery et al. 1998; Maeda et al. 1998; Maeda et al. 1999). Recent studies have shown that among the five tyrosine residues in the PIR-B cytoplasmic domain, phosphorylation of Tyr^{771} and Tyr^{801} is required for its inhibitory signal presumably through recruitment of tyrosine phosphatases SHP-1 and SHP-2(Blery et al. 1998; Maeda et al. 1998; Maeda et al. 1999). To formally demonstrate that the PIR-B cytoplasmic region containing doubly phosphorylated Tyr^{771} and Tyr^{801} is indeed the recruitment site for SHP-1 and SHP-2 in vivo, we transfected FcγRIIB-PIR-B mutant in which all remaining Tyr residues in the PIR-B cytoplasmic domain are changed to Phe (Y690, 719, 747F), into the mouse A20 B cell lymphoma IIA1.6 cells (Maeda et al. 1999). Coligation of FcγRIIB-PIR-B(Y690, 719, 747F) and BCR inhibited the calcium mobilization and recruited SHP-1 and SHP-2. Conversely, FcγRIIB-PIR-B(Y771, 810F) neither inhibited the BCR-induced calcium mobilization nor recruited SHP-1 and SHP-2. These result show that phosphorylation of remaining Tyr residues (Tyr^{690}, Tyr^{719}, Tyr^{747}) is dispensable for PIR-B-mediated inhibition and that SHP-1 and SHP-2 are recruited to the PIR-B Tyr^{771} and Tyr^{801} ITIMs in vivo. This observation, together with the evidence that both SHP-1 and SHP-2 are able to bind to the phosphorylated ITIM peptide including Tyr^{771} or Tyr^{801} in vitro (Blery et al. 1998), indicates that association of SHP-1 and SHP-2 to the PIR-B Tyr^{771} and Tyr^{801} ITIMs is mediated by SH2-phosphotyrosine dependent manner.

Lyn is required for tyrosine phosphorylation of PIR-B

The above observations prompted us to examine the responsible PTK(s) for PIR-B tyrosine phosphorylation upon its coligation with BCR. For this purpose, we obtained transformants expressing similar level of FcγRIIB-PIR-B in Lyn- or Syk-deficient mutant DT40 cells (Takata et al. 1994). Consistent with our previous results using mouse A20 B cells (Maeda et al. 1998), the chimeric receptor was slightly tyrosine phosphorylated upon BCR cross-linking alone and this phosphorylation was markedly augumented by coligation of BCR and FcγRIIB-PIR-B in wildtype DT40 B cells (Maeda et al. 1999). Syk-deficient DT40 cells showed the similar phosphorylation status of FcγRIIB-PIR-B to wildtype cells, whereas this FcγRIIB-PIR-B phosphorylation was abrogated by loss of Lyn. These results demonstrate that Lyn is required for PIR-B tyrosine phosphorylation.

Since ITAMs in Igα/Igβ are known to serve as substrates for Lyn (Reth and Wienands 1997), our results support the contention that Lyn has both activation and inhibitory roles by phosphorylating ITAMs and ITIMs, respectively (Kurosaki 1999). This inhibitory role of Lyn in B cells appears to be mediated in part by PIR-B as well as FcγRIIB and CD22 (Chan et al. 1997; Nishizumi et al. 1998; O'Rourke et al. 1997; Wang et al. 1998). Slight PIR-B tyrosine phosphorylation by BCR alone, despite less efficiently than coligation of BCR and PIR-B, might provide insights into the mechanism by which PIR-B modulates B cell activation in vivo (Ho et al 1999). Assuming that small fraction of SHP-1 or SHP-2 is recruited to the slightly phosphorylated PIR-B by BCR ligation alone, the inhibitory function of PIR-B might not necessarily require coligation of BCR and PIR-B.

PIR-B inhibits PLC-γ2 activation

IP3 generation by PLC-γ2 activation is essential for BCR-induced calcium mobilization(Sugawara et al. 1997; Takata et al. 1995). Then, Ca^{2+} enters the cytoplasm by a process termed capacitative Ca^{2+} entry, by which the depletion of intracellular stores by IP3 activates calcium influx (Putney and Bird 1993). Inhibition of the BCR-induced calcium mobilization by PIR-B could be accounted for by two possible mechanisms: inhibition of the BCR-induced IP3 generation or inhibition of capacitative Ca^{2+} entry process. BCR cross-linking alone generated IP3 while this generation was abrogated by coligation of BCR and FcγRIIB-PIR-B. Consistent with this result, BCR-induced tyrosine phosphorylation of PLC-γ2 was profoundly inhibited by its coligation with FcγRIIB-PIR-B. Thus, coligation of BCR and FcγRIIB-PIR-B inhibits PLC-γ2-mediated IP3 generation.

Genetic studies indicate the requirement of Syk and Btk for BCR-induced tyrosine phosphorylation of PLC-γ2 (Fluckiger et al. 1998; Scharenberg et al. 1998; Takata and Kurosaki 1996; Takata et al. 1994). Therefore, inhibition of PLC-γ2 activity could be explained by PIR-B-mediated dephosphorylation of PLC-γ2 and/or its upstream PTKs, Syk and Btk. The BCR-induced tyrosine phosphorylation of Syk and Btk was dramatically inhibited by coligation of BCR and FcγRIIB-PIR-B. Moreover, dephosphorylation of Syk and Btk by FcγRIIB-PIR-B resulted in inhibition of their kinase activity, assesed by their in vitro kinase assay. In addition, coligation of the BCR and FcγRIIB-PIR-B inhibited phosphorylation of both Igα and Igβ. In contrast, the tyrosine phosphorylation status and kinase activity of Lyn was not significantly changed. Thus, these observations indicate that ligation of FcγRIIB-PIR-B inhibits BCR-induced tyrosine phosphorylation of four critical signaling molecules: Igα/Igβ, Syk, Btk, and PLC-γ2. Furthermore, dephosphorylation of Syk and Btk leads to inhibition of their kinase activity.

To investigate whether the FcγRIIB-PIR-B-mediated inhibition of tyrosine phosphorylation of signaling molecules examined could be attributed to SHP-1/SHP-2, we employed vaccinia virus-driven overexpression system (Maeda et al. 1999). Expression of the catalytically inactive SHP-1-C453S, but not wildtype SHP-1, significantly reversed the FcγRIIB-PIR-B-mediated inhibition of tyrosine phosphorylation of PLC-γ2, Syk, and Btk. Moreover, kinase activity of Syk and Btk was also restored by expression of SHP-1-C453S. These results demonstrate that FcγRIIB-PIR-B ligation interrupts tyrosine phosphorylation of Syk, Btk and PLC-γ2 via a SHP-1 dependent mechnaism.

Molecular mechanism of inhibitory signal

Here we explored the molecular mechanisms of PIR-B-mediated inhibition of B cell activation. In contrast to FcγRIIB inhibition (Gupta et al. 1997; Hippen et al. 1997; Nadler et al. 1997; Ono et al. 1997), BCR-induced PLC-γ2 tyrosine phosphorylation and the subsequent IP3 generation were drastically inhibited by its coligation with PIR-B, strengthening our previous conclusion that PIR-B inhibits the B cell activation by utilizing the different inhibitory mechanism from FcγRIIB (Maeda et al. 1998). Since IP3 generation and its binding to IP3 receptors are essential for the BCR-induced calcium mobilization from intracellular pools (Sugawara et al. 1997; Takata et al. 1995), our data indicate that PIR-B aborts the initial release of Ca^{2+} from the intracellular pools by inhibiting IP3 generation. Moreover, not only PLC-γ2 but also several signaling molecules were dephosphorylated by coligation of PIR-B and BCR. Analyses of phosphotyrosine-containing proteins of whole cellular lysates also indicates the inhibition of tyrosine phosphorylation of many proteins by PIR-B (data not shown). Therefore, our data favor the possibility that upstream PTKs rather than PLC-γ2 itself are down-modulated by PIR-B ligation.

Early activation of PTKs, particularly Syk and Btk, is involved in PLC-γ2 activation upon BCR engagement. Indeed, BCR-induced phosphorylation and activation of Syk and Btk was profoudly inhibited by coligation of PIR-B and BCR, and this inhibition was substantially recovered by overexpression of SHP-1-C453S. PIR-B-mediated dephosphorylation of Igα/Igβ also could account for down-modulation of Syk (Campbell 1999; Kurosaki 1999; Reth and Wienands 1997). The other hand, tyrosine phosphorylation and activity of Lyn was not significantly inhibited by PIR-B ligation. Together with the previous results that

PIR-B-mediated inhibition requires redundant functions of SHP-1 and SHP-2 (Maeda et al. 1998), our data indicate that among three types of BCR-activated PTKs, Syk and Btk are specifically down-modulated by SHP-1 and SHP-2, thereby leading to inhibition of PLC-γ2 phosphorylation.

Based on these results, we propose following mechanism: (a) Coligation of PIR-B and BCR results in phosphorylation of PIR-B ITIMs by Lyn; (b) SHP-1 and SHP-2 are recruited to the phosphorylated PIR-B ITIMs via their SH2 domains, and are subsequently activated; (c) activated SHP-1 and possibly SHP-2 dephosphorylate their substrates apparently including Syk and Btk; (d) down-modulation of Syk and Btk by presumably dephosphorylation of tyrosine residues in their activation loop inhibits PLC-γ2 activation; (e) down-modulated PLC-γ2 decreases inositol 1,4,5-trisphosphate generation leading to a block of calcium mobilization (Fig.1).

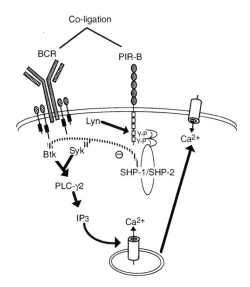

Fig. 1. Mechanism of inhibitory signaling mediated by PIR-B. Co-ligation of PIR-B promotes tyrosine phosphorylation of ITIMs of PIR-B cytoplasmic domain, providing binding sites to SH2 domains of SHP-1/SHP-2, which in turn dephosphorylate Syk and Btk. These dephosphorylated Syk and Btk result in downregulation of PLC-γ2, leading to inhibition of calcium release.

References

Bléry M, Kubagawa H, Chen CC, Vély F, Cooper MD, Vivier E (1998) The paired Ig-like receptor PIR-B is an inhibitory receptor that recruits the protein-tyrosine phosphatase SHP-1. Proc Natl Acad Sci USA 95:2446-2451

Campbell KS (1999) Signal transduction from the B cell antigen-receptor. Curr Opin Immunol 11:256-264

Chan VWF, Meng F, Soriano P, DeFranco AL, Lowell CA (1997) Characterization of the B lymphocyte populations in Lyn-deficient mice and the role of Lyn in signal initiation and down-regulation. Immunity 7:69-81

Fluckiger A-C, Li Z, Kato RM, Wahl MI, Ochs HD, Longnecker R, Kinet J-P, Witte ON, Scharenberg AM, Rawlings DJ (1998) Btk/Tec kinases regulate sustained increases in intracellular Ca^{2+} following B-cell receptor activation. EMBO J 17:1973-1985

Gupta N, Scharenberg AM, Burshtyn DN, Wagtmann N, Lioubin MN, Rohrschneider LR, Kinet J-P, Long EO (1997) Negative signaling pathways of the killer cell inhibitory receptor and FcγRIIb1 require distinct phosphatases. J Exp Med 186:473-478

Hayami K, Fukuta D, Nishikawa Y, Yamashita Y, Inui M, Ohyama Y, Hikida M, Ohmori H, Takai T (1997) Molecular cloning of a novel murine cell-surface glycoprotein homologous to killer cell inhibitory receptors. J Biol Chem 272:7320-7327

Hippen KL, Buhl AM, D'AmbrosioD, Nakamura K, Persin C, Cambier JC (1997) FcγRIIB1 inhibition of BCR-mediated phosphoinositide hydrolysis and Ca^{2+} mobilization is integrated by CD19 dephosphorylation. Immunity 7:49-58

Ho LH, Uehara T, Chen CC, Kubagawa H, Cooper MD (1999) Constitutive tyrosine phosphorylation of the inhibitory paired Ig-like receptor PIR-B. Proc Natl Acad Sci USA 96:15086-15090

Kubagawa H, Burrows PD, Cooper MD (1997) A novel pair of immunoglobulin-like receptors expressed by B cells and myeloid cells. Proc Natl Acad Sci USA 94:5261-5266

Kurosaki T (1999) Genetic analysis of B cell antigen receptor signaling. Annu Rev Immunol 17:555-592

Maeda A, Kurosaki M, Ono M, Takai T, Kurosaki T (1998) Requirement of tyrosine phosphatase SHP-1 and SHP-2 for PIR-B-mediated inhibitory signal. J Exp Med 187:1355-1360

Maeda A, Scharenberg AM, Tsukada S, Bolen JB, Kinet J-P, Kurosaki T (1999) PIR-B inhibits BCR-induced activation of Syk and Btk by SHP-1. Oncogene 18:2291-2297

Nadler MJS, Chen B, Anderson JS, Wortis HH, Neel BG (1997) Protein-tyrosine phosphatase SHP-1 is dispensable for FcγRIIB-mediated inhibition of B cell antigen receptor activation. J Biol Chem 272:20038-20043

Nishizumi H, Horikawa K, Mlinaric-Rascan I, Yamamoto, T (1998) A double-edged kinase Lyn: a positive and negative regulator for antigen receptor-mediated signals. J Exp Med 187:1343-1348

Ono M, Okada H, Bolland S, Yanagi S, Kurosaki T, Ravetch JV (1997) Deletion of SHIP or SHP-1 reveals two distinct pathways for inhibitory signaling. Cell 90:293-301

O'Rourke L, Tooze R, Fearon DT (1997) Co-receptors of B lymphocytes. Curr Opin Immunol 9:324-329

Putney JWJr, Bird GSJ (1993) The signal for capacitative calcium entry. Cell 75:199-201

Reth M, Wienands J (1997) Initiation and processing of signals from the B cell antigen receptor. Annu Rev Immunol 15:453-479

Scharenberg AM, El-Hillal O, Fruman DA, Beitz LO, Li Z, Lin S, Gout I, Cantley LC, Rawlings DJ, Kinet J-P (1998) Phosphatidylinositol-3,4,5-trisphosphate(Ptdlin-3,4,5-P3)/Tec kinase-dependent calcium signaling pathway: a target for SHIP-mediated inhibitory. EMBO J 17:1961-1972.

Sugawara, H, Kurosaki M, Takata M, Kurosaki T (1997) Genetic evidence for involvement of type 1, type 2, and type 3 inositol 1,4,5-trisphosphate receptors in signal transduction through the B cell antigen receptor. EMBO J 16:3078-3088

Takata M, Homma Y, Kurosaki T (1995) Requirement of phospholipase C-γ2 activation in surface immunoglobulin M-induced B cell apoptosis. J Exp Med 182:907-914

Takata M, Kurosaki T. (1996) A role for Bruton's tyrosine kinase in BCR-mediated activation of phospholipase C-γ2. J Exp Med 184:31-40

Takata M, Sabe H, Hata A, Inazu T, Homma Y, Nukada T, Yamamura H, Kurosaki T (1994) Tyrosine kinases Lyn and Syk regulate B cell receptor-coupled Ca^{2+} mobilization through distinct pathways. EMBO J 13:1341-1349

Wang J, Koizumi T, Watanabe T (1996) Altered antigen receptor signaling and impaired fas-mediated apoptosis of B cells in Lyn-deficient mice. J Exp Med 184:831-838

23

IgE stabilizes its high affinity receptor (FcεRI) on mast cells in vitro and ex vivo: The mechanism of IgE-mediated FcεRI up-regulation and its physiological meaning

S. Kubo[1], K. Matsuoka[1, 2], C. Taya[3], F. Kitamura[1], H. Yonekawa[3], and H. Karasuyama[1,4]

[1]Department of Immunology, The Tokyo Metropolitan Institute of Medical Science, Bunkyo-ku, Tokyo 113-8613, Japan
[2]Japan Science and Technology Corporation, Kawaguchi, 332-0012, Japan
[3]Department of Laboratory Animal Science, The Tokyo Metropolitan Institute of Medical Science, Bunkyo-ku, Tokyo 113-8613, Japan
[4]Department of Immune Regulation, Tokyo Medical and Dental University, Gradual School, Bunkyo-ku, Tokyo 113-8519, Japan

Summary. The interaction of IgE with the high affinity receptor for immunoglobulin E (designated FcεRI) is central to IgE dependent anaphylaxis. Cross-linking of FcεRI initiates an intracellular signal transduction cascade that triggers the release of mediators of the allergic response. Surface FcεRI levels of mast cells and basophils are regulated by IgE. There is a correlation between the number of FcεRI and the severity of effector responses. Ligand-mediated FcεRI up-regulation should be one of the key mechanisms to control IgE response. Here we study the major mechanism by which IgE mediates FcεRI up-regulation. Intracellular protein transport inhibitor Brefeldin A (BFA) inhibited IgE-mediated FcεRI up-regulation. In the presence of BFA, IgE-free FcεRI disappeared from cell surface rapidly, whereas IgE-bound FcεRI were lost very slowly. Total amount of new FcεRI expression did not depend on total number of IgE-bound FcεRI. In human FcεRIα transgenic mice, human IgE enhanced expression of human FcεRI, but mouse FcεRI levels were the same as before addition of human IgE. These suggest that stabilization of FcεRI by IgE is the major mechanism of IgE-mediated FcεRI up-regulation.

Key words. IgE, FcεRI, Allergy, Mast cell

Regulation of FcεRl expression

The immunoglobulin E (IgE)-mediated allergic response requires the binding of IgE to its high-affinity receptor FcεRI. Cross-linking of FcεRI initiates an intracellular signal transduction cascade that triggers the release of mediators of the allergic response. The interaction of the IgE with FcεRI is central to these immune reactions (Kinet 1999; Turner and Kinet 2000).

FcεRI is composed of three different chains α, β and γ. Human FcεRI can be expressed as either $\alpha\gamma_2$ or $\alpha\beta\gamma_2$. Mast cells and basophils express the $\alpha\beta\gamma_2$ tetramer, whereas monocytes, cutaneous Langerhans cell express only the $\alpha\gamma_2$ trimer. The α chain binds the Fc portion of IgE with high affinity. The β and γ chains are involved in signal transduction.

One of the major mechanisms of regulation of surface FcεRI expression is IgE-mediated FcεRI up-regulation. In 1978, a correlation between the number of the IgE molecules bound to basophils and the serum IgE level was reported (Malveaux et al. 1978). Furthermore, it was demonstrated, incubation of rat mast cell line RBL-2H3 cells with monomeric IgE for 12 to 24 h resulted in a mild increase in the number of receptor at the cell surface (Furuichi et al. 1985; Quatro et al. 1985). In late 1990, more direct evidence of IgE-mediated FcεRI up-regulation was reported. In IgE deficient mice, the level of FcεRI expression on peritoneal mast cells was famine to be extremely low. The level of surface expression of FcεRI on the bone marrow derived mast cells (BMMCs) was up-regulated (up to 32-folds) by in vitro incubation with IgE (Yamaguchi et al. 1997). IgE-mediated FcεRI up-regulation was observed on mouse basophils (Lantz et al. 1977) and human mast cells (Yano et al. 1997), and basophils (MacGlashan et al. 1998).

The current study addressed the mechanism of IgE-mediated FcεRI up-regulation. Two mechanisms have been discussed: increased synthesis and decreased catabolism. In an early study with RBL-2H3, slow disappearance of IgE-bound FcεRI and rapid disappearance of IgE-free FcεRI on the cell surface were demonstrated, but extent of FcεRI up-regulation was very limited (two to three folds) (Furuichi et al. 1985; Quatro et al. 1985). In contrast, much more FcεRI up-regulation mediated by IgE was observed in vivo and in vitro (20 to 30 folds up-regulation) in mouse and human. Therefore, it is not clear whether such drastic up-regulation of FcεRI can be explained by the stabilization of IgE-bound FcεRI. It could be mainly due to the enhancement of FcεRI synthesis and transport cells surface. To address this issue, we examined the turnover of FcεRI on BMMCs and peritoneal mast cells.

Enhanced FcεRI expression in IgE transgenic mice

We have recently established IgE transgenic mice which constitutivly produce hapten TNP-specific IgE (Matsuoka et al. 1999). Their sera contain 30∼50μg/ml of IgE while non-transgenic littermates' sera contain only 1 μg/ml of IgE. The transgenic mice help to understand the immune system under high concentration of IgE. For example, low affinity receptor of IgE, CD23, on splenic B cells is up-regulated up to five times in IgE transgenic mice compared to non-transgenic littermate mice. IgE-bound CD23 on B cells in IgE transgenic mice is also 40 folds higher than that of normal mice. To understand IgE-mediated FcεRI up-regulation in vivo, surface levels of FcεRI on c-kit$^+$ peritoneal mast cells were examined by using three different mouse strains B-cell deficient mice μm-/-, BALB/c and IgE transgenic mice. Surface FcεRI expression levels of peritoneal mast cells from the IgE transgenic mice were five times as high as those from BALB/c mice. On the other hand, FcεRI expression levels from μm-/- mice were ∼25% of these on mast cells from normal BALB/c mice just as observed in IgE-/- mice (Yamaguchi et al. 1997). These observations indicate that the levels of FcεRI on mast cells can be altered in vivo at least by 20-fold in correlation with serum IgE levels and IgE binding.

IgE-bound FcεRI has much stronger stability than IgE-free FcεRI on both BMMCs and peritoneal mast cells

BFA, one of the protein transfer blocker, completely inhibited the transport of new FcεRI to the cells surface and IgE-mediated FcεRI up-regulation on BMMCs in vitro. Thus, in the presence of BFA, we could chase fate of surface FcεRI as the absence of newly transported FcεRI. IgE-free FcεRI disappeared very rapidly from the cell surface of BMMCs. During 12h culture, 70% of IgE-free FcεRI was lost, whereas only 10% of IgE-bound FcεRI was lost from the cell surface. Using peritoneal mast cells, we obtained the same results. These experiments showed strong stability of IgE-bound FcεRI and unstability of IgE-free FcεRI.

We next compared the extent of up-regulation of FcεRI in two different conditions: One in which levels of FcεRI expression was low, the other in which those were already up-regulated by pre-incubation with IgE. Incubation with IgE induced the comparable levels of increase in FcεRI expression in both conditions even though total numbers of FcεRI were quite different between the two conditions. This suggests that the rate of FcεRI transport is constant regardless of FcεRI levels. Therefore it is unlikely that IgE binding to FcεRI promotes the acceleration of FcεRI synthesis and/or transport to the cell surface.

In human FcεRIα transgenic mice, mast cells express human FcεRIα and mouse FcεRIα. Both chains share mouse β and γ chains involved in signal transduction. Incubation of BMMCs from the transgenic mice with human IgE induced drastic increase of human chimeric FcεRI, but did not give any significant effect on the expression of mouse FcεRI. These observations support that the major mechanism of IgE-mediated FcεRI up-regulation can be stabilization of FcεRI through binding IgE to FcεRI.

Acquisition of specificity and formation of short memory through ligand-mediated FcεRI up-regulation: novel view of mast cell

What is physiological meaning of IgE-mediated FcεRI up-regulation? Drastic up-regulation of FcεRI may be a critical mechanism to enhance host defense against various types of parasites. Indeed, it has been shown that IgE-mediated FcεRI up-regulation is important for production of IL-4 (Yamaguchi et al. 1997).

In acquired immunity, specificity and maintenance of reactivity (memory) are essential functions. T and B cells acquire antigen specificity through rearrangements of antigen receptor genes. Therefore, each cell shows a unique specificity. On the other hand, mast cells depend on its FcεRI to acquire antigen specificity. IgE bound to FcεRI gives antigen specificity to mast cells. In healthy BALB/c mice, 80% of FcεRI are occupied with polyclonal IgE. Therefore, if no up-regulation of FcεRI occurs, only 20% of FcεRI are available for newly produced IgE against pathogens in case of infection. IgE-mediated FcεRI up-regulation overcomes this limitation of FcεRI. The amount of newly produced IgE-bound FcεRI increases drastically. This is alternative way to acquire antigen specificity in immune system (Fig. 1).

T cells have long memory and give a qualitatively different and quantitatively enhanced response. Memory of mast cells depends on stability of surface IgE-bound FcεRI and the number of IgE-bound FcεRI to maintain reactivity. During IgE production, the number of FcεRI increase and accumulation FcεRI loaded with newly produced IgE occurs. In other word, this is the formation of short memory of mast cells. Acquisition of specificity and formation of short memory through ligand-mediated FcεRI up-regulation are two sides of the same coin. Those mast cells can react to the pathogens without proliferation. This short memory may help protection from short-term repeated re-infections (Hagan et al. 1991).

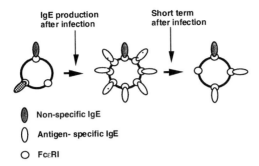

Fig. 1. Acquisition of antigen specificity and short memory formation through IgE-mediated FcεRI up-regulation. In healthy condition (BALB/c), 80% of FcεRI receptors on mast cell are associated with IgE. Only limited numbers of FcεRI are available for newly produced antigen specific IgE. When parasite infections and exposure of many allergens cause IgE production, the number of FcεRI is up-regulated drastically by IgE-mediated FcεRI regulation. This enhancement of FcεRI expression makes newly produced IgE possible to bind to mast cells, resulting on the acquisition of antigen specificity. Since, half-life of IgE-bound FcεRI is much longer than IgE-free FcεRI, mast cells can keep the antigen specificity for a while even after serum levels of the antigen-specific IgE goes down.

References

Furuichi K, Rivera J, Isersky C (1985) The receptor for immunoglobulin E on rat basophilic leukemia cells: Effect of ligand binding on receptor expression. Proc Natl Acad Sci USA 82:1522-1525

Hagan P, Blumenthal UJ, Dunn D, Simpson AJG, Wilkins HA (1991) Human IgE, IgG4 and resistance to reinfection with Schistosoma haematobium. Nature 349:243-245

Kinet J-P (1999) The high-affinity IgE receptor (FcεRI): From physiology to pathology. Annu Rev Immunol 17:931-972

Lantz CS, Yamaguchi M, Oettgen HC, Katona IM, Miyajima I, Kinet J-P, Galli SJ (1997) IgE regulates mouse basophil FcεRI expression in vivo. J Immunol 158:2517-2521

MacGlashan D Jr, McKenzie-White J, Chichester K, Bochner BS, Davis FM, Schroeder JT, Lichtenstein LM (1998) In vitro regulation of FcεRIα expression on human basophils by IgE antibody. Blood. 91:1633-1643

Malveaux FJ, Conroy MC, Adkinson NF Jr, Lichtenstein LM (1978) IgE receptors on human basophils: Relationship to serum IgE concentration. J Clin Invest 62:176-181

Matsuoka K, Taya C, Kubo S, Toyama-Sorimachi N, Kitamura F, Ra C, Yonekawa H, Karasuyama H (1999) Establishment of antigen-specific IgE transgenic mice to study pathological and immunobiological roles of IgE in vivo. Int Immunol 11:987-994

Turner H, Kinet J-P (1999) Signalling through the high-affinity IgE receptor FcεRI. Nature 402:B24-B30

Quarto R, Kinet J-P, Metzger H (1985) Coordinate synthesis and degradation of the α-, β- and γ-subunits of the receptor for immunoglobulin E. Mol Immunol 22:1045-1051

Yamaguchi M, Lantz CS, Oettgen HC, Katona IM, Fleming T, Miyajima I, Kinet J-P, Galli SJ (1997) IgE enhances mouse mast cell FcεRI expression in vitro and in vivo: Evidence for a novel amplification mechanism in IgE-dependent reactions. J Exp Med 185:663-672

Yano K, Yamaguchi M, de Mora F, Lantz CS, Butterfield JH, Costa JJ, Galli SJ (1997) Production of macrophage inflammatory protein-1α by human mast cells: Increased anti-IgE-dependent secretion after IgE-dependent enhancement of mast cell IgE-binding ability. Lab Invest 77:185-193

24
Autoimmune arthritis and Goodpasture's syndrome induced in Fcγ receptor-deficient mice

Toshiyuki Takai and Akira Nakamura

Department of Experimental Immunology and CREST Program of JST, Institute of
Development, Aging and Cancer, Tohoku University, Seiryo 4-1, Sendai 980-8575, Japan

Summary. To address the question whether the type IIB Fc receptor for IgG
(FcγRIIB)-deficient mice (RIIB$^{-/-}$) are susceptible to induction of various
autoimmune diseases, we tested to induce type II collagen-induced arthritis (CIA),
a model for rheumatoid arthritis in humans. We found that RIIB$^{-/-}$ on a non-
permissive H-2^b background become susceptible to CIA induction. Moreover, we
could induce Goodpasture's syndrome (GPS) in RIIB$^{-/-}$ by immunization with
type IV collagen. Quite similar to human GPS, RIIB$^{-/-}$ develop massive
pulmonary hemorrhage and glomerulonephritis. These results highlight the role of
FcγRIIB in maintaining tolerance and suggest that it may play a critical role in the
pathogenesis of rheumatoid arthritis and GPS in humans.

Key words. Collagen-induced arthritis, Autoimmune glomerulonephritis, Fc
receptor, Goodpasture's syndrome, Gene targeting

Introduction

Triggering the activation receptors, FcγRI and FcγRIII or FcεRI elicit a variety of
effector functions, including phagocytosis, antibody-dependent cell-mediated
cytotoxicity, and the release of inflammatory mediators (reviewed in Ravetch and
Kinet 1991; Metzger 1992). Analysis of FcR-deficient mice has revealed that
these receptors play in the mechanism of initiating type I, II, and type III
hypersensitivity reactions and anaphylaxis (Hazenbos et al. 1996; Takai et al.
1994; Sylvestre and Ravetch 1994; Clynes and Ravetch 1995; Dombrowicz et al.
1993; Miyajima et al. 1997; Dombrowicz et al. 1997; Clynes et al. 1998; Suzuki et
al. 1998; Wakayama et al. 2000). On the other hand, FcγRIIB, the most widely
expressed FcR, inhibits cellular activation triggered through antibody or immune
complexes (Fridman et al. 1992; van de Winkel and Capel 1993; Ravetch 1994;
Daëron et al. 1995; Takai and Ravetch 1998). Disruption of FcγRIIB by gene

targeting resulted in mice with elevated immunoglobulin levels in response to both thymus-dependent and thymus-independent antigens, enhanced passive cutaneous and systemic anaphylaxis reactions (Takai et al. 1996; Ujike et al. 1999), and enhanced immune complex mediated inflammatory responses (Clynes et al. 1999; Suzuki et al. 1998; Schiller et al. 2000). These studies indicate that FcγRIIB physiologically acts as a negative regulator of immune complex-triggered activation and may function in vivo to suppress autoimmunity by regulating both B cell responses and effector cell activation. However, RIIB$^{-/-}$ do not develop spontaneous autoimmune diseases at least on a hybrid 129/Sv/C57BL/6 background under specific pathogen-free condition.

Collagen-induced arthritis in RIIB$^{-/-}$ mice

CIA, a model for rheumatoid arthritis in humans, is a chronic inflammatory arthropathy that can be induced in susceptible rodents by immunization with C-II (Courtenay et al. 1980; Holmdahl et al. 1985; Trentham et al. 1977). The development of arthritis is thought to be associated with the synergistic effect of high levels of cell-mediated and humoral immunity to C-II. CIA and rheumatoid arthritis are clearly associated with the MHC region (Nabozny et al. 1996), and in mice only H-2^q and H-2^r haplotypes are susceptible to CIA (Wooley et al. 1981). The responsible gene in the H-2^q haplotype has been identified as the A^q class II molecule (Brunsberg et al. 1994), which binds peptides derived from C-II, thus leading to T cell activation which is of crucial importance for development of arthritis in this model. In addition, a strong B cell response is activated in CIA, producing IgG directed towards C-II-specific structures. There is evidence that these antibodies are directly pathogenic, as shown by transfer experiments (Stuart and Dixon 1983; Holmdahl et al. 1990), as well as synergizing with activated T cells to promote the development of arthritis (Holmdahl et al. 1994; Seki et al. 1988). B cell-deficient mice on a susceptible background do not develop CIA, indicating that B cells play crucial role for development of CIA (Svensson et al. 1998).

RIIB$^{-/-}$ were generated (Takai et al. 1996) on 129/Sv (H-2^b) and C57BL/6 (H-2^b) hybrid genetic background, a non-susceptible haplotype. However, we found that RIIB$^{-/-}$ mice develop arthritis with a time course and severity comparable to DBA/1 mice when immunized with C-II (Yuasa et al. 1999) (Table 1). While the incidence of arthritis in RIIB$^{-/-}$ was lower than DBA/1, it was dramatically enhanced compared to wild-type H-2^b mice. In those mice that developed arthritis, the mean onset of disease for RIIB$^{-/-}$ was comparable to that in DBA/1 controls. Similarly the mean maximal arthritic index of the mutant animals was also comparable to DBA/1 controls. Histopathological features of the CIA induced in RIIB$^{-/-}$ revealed that the arthritic lesions of the RIIB$^{-/-}$ showed massive lymphocytic and monocyte/macrophage infiltration associated with cartilage-bone destruction similar to that observed in DBA/1 immunized animals. Thus, RIIB$^{-/-}$

immunized with C-II developed a destructive arthritis, which is qualitatively similar to those of arthritis in DBA/1 mice.

Table 1. Summary of the CIA course in RIIB[-/-] mice

Mice	No.	Incidence[a] (%)	Onset[b] (days)	Arthritic index[c]	Arthritic paws[d] (%)
Wild type	43	7.0	50.5 ± 7.6	2.3 ± 1.9	4.1
RIIB[-/-]	45	42.2	35.3 ± 12.5	6.9 ± 3.6	26.7
DBA/1	42	95.2	33.2 ± 8.6	8.6 ± 1.9	67.3

[a] $P < 0.001$ for RIIB-/- *vs* wild type, and RIIB-/- *vs* DBA/1.
[b] $P < 0.05$ for RIIB-/- *vs* wild type, not significant for RIIB-/- *vs* DBA/1.
[c] $P < 0.05$ for RIIB-/- *vs* wild type, not significant for RIIB-/- *vs* DBA/1.
[d] $P < 0.001$ for RIIB-/- *vs* wild type, and RIIB-/- *vs* DBA/1.

Antibodies specific for C-II play a major role in the pathogenesis of CIA (Holmdahl et al. 1985, Stuart and Dixon 1983). As expected, RIIB[-/-] had higher anti-C-II antibody titers than those of wild-type mice in all isotypes tested. The anti-C-II IgG antibody response is enhanced in RIIB[-/-], especially in those mice exhibiting arthritis, suggesting that the relatively high anti-C-II IgG level could be one of the pathogenic factors, although unlikely by itself to explain the induction of disease in the H-2[b] background.

Since CIA is dependent upon dysregulation of both humoral and cell-mediated responses, we next determined whether the absence of FcγRIIB altered the phenotype of the cell-mediated immune response to C-II. We compared the specific proliferative responses and cytokine production of C-II-primed lymph node cells derived from RIIB[-/-], wild-type H-2[b] and DBA/1, but we found that the response in RIIB[-/-] was much lower than DBA/1 control, indicating that disruption of FcγRIIB does not appreciably modify the antigen specific T cell response in non-permissive animals and is not likely to account for the susceptibility of these animals to CIA.

At later stages of autoimmune arthritis, local synthesis of cytokines and other inflammatory mediators are likely to be responsible for the progression from inflammation to a destructive arthritis. To determine if macrophages derived from RIIB[-/-] showed enhanced release of inflammatory mediators upon stimulation, we measured the levels of IL-1α produced upon stimulation with IgG-opsonized sheet red blood cells. We found thioglycollate-elicited peritoneal macrophages from RIIB[-/-] released IL-1α more abundantly than those from wild-type controls and at levels comparable to macrophages derived from DBA/1 mice. Thus, the lack of FcγRIIB makes macrophages more sensitive to stimulation with IgG-immune complexes, and results in a higher level of secretion of the proinflammatory mediator.

Recently, Kleinau et al. (2000) generated RIIB$^{-/-}$ and FcR common γ chain-deficient mice (FcRγ$^{/-}$) on DBA/1 background. FcRγ$^{/-}$/DBA were protected from CIA, whereas RIIB$^{-/-}$/DBA developed an augmented IgG anti-collagen response and arthritis. Their observations suggest a crucial role of FcγRI and FcγRIII in triggering autoimmune arthritis, in addition to an inhibitory role of FcγRIIB described above. Moreover, it shoud be noted that Bolland and Ravetch (2000) found very recently that RIIB$^{-/-}$ on C57BL/6 genetic background, but not on BALB/c background, develop autoantibodies and autoimmune glomerulonephritis. Transfer experiment indicated that the autoimmune phenotype is associated with the presence of donor RIIB-deficient cells, with the RIIB-sufficient myeloid cells primarily derived from the recipient, suggesting that FcγRIIB deficiency on B cells leads to autoimmune disease in specific genetic backgrounds.

Goodpasture's syndrome in RIIB$^{-/-}$ mice

GPS is characterized by a rapid and progressive glomerulonephritis and hemorrhagic pneumonitis with often fatal results (reviewed in Goodpasture 1919; Kalluri 1999). The presence of antibodies to type IV collagen (C-IV) antibodies and the immune complex deposition along the basement membranes of both lung and glomeruli has lead to a proposed pathogenic model for this disease (Lener 1987; Gunwar 1991; Savage et al. 1996). The mechanism(s) which result in the loss of tolerance and the development of autoantibodies to C-IV are unknown. Importantly, investigations into the proposed pathogenesis of GPS and development of therapeutic strategies have been hampered by the lack of a suitable animal model. Attempts to develop such a model for GPS, by immunizing sheep (Steblay 1979) or rats (Couser et al. 1985; Abate et al. 1998) with homologous or heterologous glomerular basement membranes preparations have resulted in only partial success. Based on our observations in CIA model in RIIB$^{-/-}$ (Yuasa et al. 1999), we wanted to ask whether FcγRIIB may regulate responses to C-IV and be a critical factor in GPS development. Our data indicate that immunization of RIIB$^{-/-}$ with C-IV results in autoimmunity and culminates in GPS-like disease (Table 2), suggesting a role for the FcγRIIB regulatory pathway in the etiology of this autoimmune disease (Nakamura et al. 2000).

RIIB$^{-/-}$, immunized with bovine C-IV in CFA and boosted 3 times at every 2 weeks with antigen in IFA, developed a GPS-like disease with pulmonary hemorrhage and glomerulonephritis (Nakamura et al. 2000) (Table 2). Nearly all immunized RIIB$^{-/-}$ developed pulmonary hemorrhage while none of the wild-type controls nor FcRγ$^{/-}$ display evidence of disease. The appearance of the diseased lungs revealed extensive hemorrhage, in sharp contrast to the results reported in previous attempts to induce GPS in rodents (Couser et al. 1985; Abate et al. 1998; Kalluri et al. 1997), in which only regional hemorrhagic patches have been observed.

Table 2. Summary of GPS in the mice immunized with C-IV

Mice	No.	% LH	% GN[a]	% CG	% Mort.	BUN[c] (mg/dl)	Cr[c] (mg/dl)
RIIB$^{-/-}$	24	79	92	33.8 ± 25.0	21[b]	45.7 ± 16.0	0.15 ± 0.06
Wt	16	0	0	0	0	23.3 ± 8.6	0.06 ± 0.02
FcR$\gamma^{-/-}$	14	0	0	0	0	23.4 ± 8.6	0.07 ± 0.03

[a]All mice with glomerulonephritis had tubulointerstitial nephritis.
[b]Five mice died of the tracheal obstruction caused by hemoptysis.
[c]$P < 0.01$ for RIIB$^{-/-}$ vs wild-type mice.
LH, lung hemorrhage; GN, glomerulonephritis; CG, crescentic glomeruli; Mort, mortality; BUN, blood urine nitrogen; Cr, creatinine.

Histopathological examination of the kidneys in immunized RIIB$^{-/-}$ revealed typical crescentic glomerulonephritis and tubulointerstitial nephritis, with infiltration of neutrophils and multinuclear giant cells, and hyaline droplet degeneration of the tubules in nearly all mice tested (Table 2), whereas those from wild-type mice did not show any pathological changes. Immunized RIIB$^{-/-}$ showed higher blood urea nitrogen and serum creatinine levels than controls (Table 2), further confirming the compromised renal function. In addition, all diseased animals showed proteinuria with mean protein levels in the urine of 533 mg/dl. Immunohistochemical staining of glomeruli from immunized RIIB$^{-/-}$ with FITC-labeled anti-mouse IgG demonstrated linear depositions of IgG-immune complexes along the glomerular and tubulointerstitial membranes. In contrast, the deposition of IgG was hardly detectable in wild-type or FcR$\gamma^{-/-}$, immunized mice. The identification of FcγRIIB as a susceptibility factor in GPS and the ability to induce disease upon immunization with bovine C-IV offers the opportunity to investigate other presumptive triggers in the RIIB$^{-/-}$ such toxic oxygen, hydrocarbons, or viruses, all of which are suggested as possible inducers of human GPS (Kelly and Haponik 1994).

The pulmonary and renal lesions of GPS are attributed to anti-glomerular basement membrane antibodies, which bind to common antigenic sites in the lung and kidney and activate inflammatory effector responses. Consistent with this pathogenic model, the antibody response of RIIB$^{-/-}$ to bovine C-IV immunization is enhanced relative to wild-type controls, with elevated titers of anti-bovine C-IV antibodies. Importantly, RIIB$^{-/-}$, but not wild-type or FcR$\gamma^{/-}$, showed elevated autoantibody responses to mouse C-IV, suggesting that murine GPS depends on the autoantibody production in this susceptible background. While this enhanced antibody response may account for the GPS disease which develops in these animals, it is likely that contributions from enhanced effector cell responses to deposited anti-C-IV autoantibodies are a significant factor in the development of disease as described in the immune complex-mediated alveolitis (Clynes et al. 1999) and in CIA (Yuasa et al. 1999). In support of this notion, the alveolar and glomerular diseases could be transferred only to RIIB$^{-/-}$ but not to wild-type mice by injecting CIV-sensitized splenocytes from RIIB$^{-/-}$ to non-sensitized animals. Disease could not be transferred by injecting either B220$^+$ or B220$^-$ splenocytes from C-IV-sensitized RIIB$^{-/-}$ into non-sensitized RIIB$^{-/-}$, suggesting the additional requirement of sensitized T and B cell co-operation in disease development (Nakamura et al. 2000).

Conclusion

Studies on CIA in RIIB$^{-/-}$ suggest that the development of the autoimmune collagen arthritis represents the dysregulation of both humoral and effector pathways. FcγRIIB is a pleiotropic receptor, functioning to down-regulate both B cell and effector cell responses. Analysis of CIA in RIIB$^{-/-}$ on DBA background further supports this notion. Similar role of FcγRIIB may be found in other autoimmune disease models and in human susceptibility to autoimmune disease. This speculation has been supported by our findings in another autoimmune disease, GPS, induced in RIIB$^{-/-}$. The data on the role of FcγRIIB in GPS as well as in CIA suggest that alterations in the function or expression of this gene could be a susceptibility factor in the pathogenesis of the human diseases. Strategies which result in the up-regulation of this receptor and its signaling would therefore represent potential new therapeutic approaches to the treatment of autoimmune diseases.

References

Abate M, Kalluri R, Corona D, Yamaguchi N, McCluskey RT, Hudson BG, Andres G, Zoja C, Remuzzi G (1998) Experimental Goodpasture's syndrome in Wistar-Kyoto rats immunized with α3 (IV) chain of the type IV collagen. Kidney Int 54:1550-1561

Bolland S, Ravetch JV (2000) Spontaneous autoimmune disease in FcγRIIB-deficient mice results from strain-specific epistasis. Immunity 13:277-285

Brunsberg U, Gustafson K, Jansson L, Micha Isson E, hrlund-Richter L, Pettersson S, Mattsson R, Holmdahl R (1994) Expression of transgenic class II A^b gene confers susceptibility to collagen-induced arthritis. Eur J Immunol 24:1698-702

Clynes R, Dumitru C, Ravetch JV (1998) Uncoupling of immune complex formation and kidney damage in autoimmune glomerulonephritis. Science 279:1052-1054

Clynes R, Maizes JS, Guinamard R, Ono M, Takai T, Ravetch JV (1999) Modulation of immune complex induced inflammation in vivo by the coordinate expression of activation and inhibitory Fc receptors. J Exp Med 189:179-186

Clynes R, Ravetch JV (1995) Cytotoxic antibodies trigger inflammation through Fc receptors. Immunity 3:21-26

Courtenay JS, Dallman MJ, Dayan AD, Marten A, Mosedale B (1980) Immunization against heterologous type II collagen induces arthritis in mice. Nature 283:666-668

Couser WG, Darby C, Salant DJ, Adler S, Stilmant MM, Lowenstein LM (1985) Anti-GBM antibody-induced proteinuria in isolated perfused rat kidney. Am J Physiol 249:F241-F250

Daëron M, Latour S, Malbec O, Espinosa E, Pina P, Pasmans S, Fridman WH (1995) The same tyrosine-based inhibition motif, in the intracytoplasmic domain of FcgRIIB, regulates negatively BCR-, TCR-, and FcR-dependent cell activation. Immunity 3:635-646

Dombrowicz D, Flamand V, Brigman KK, Koller BH, Kinet J-P (1993) Abolition of anaphylaxis by targeted disruption of the high affinity immunoglobulin E receptor α chain gene. Cell 75:969-976

Dombrowicz D, Flamand V, Miyajima I, Ravetch JV, Galli SJ, Kinet J-P (1997) Absence of FcεRI α chain results in upregulation of FcγRIII-dependent mast cell degranulation and anaphylaxis. Evidence of competition between FcεRI and FcγRIII for limiting amounts of FcR β and γ chain. J Clin Invest 99:915-925

Fridman WH, Bonnerot C, Daëron M, Amigorena S, Teillaud JL, Sautes C (1992) Structural bases of Fcγ receptor functions. Immunol Rev 125:49-76

Goodpasture EW (1919) The significance of certain pulmonary lesions in relation at the etiology of influenza. Am J Med Sci 158:863-870

Gunwar S (1991) Alveolar basement membrane: Molecular properties of the noncollagenous domain (hexamer) of collagen IV and its reactivity with Goodpasture autoantibodies. Am J Respir Cell Mol Biol 5:107-112

Hazenbos WLW, Gessner JE, Hofhuis FMA, Kuipers H, Meyer D, Heijnen IAFM, Schmidt RE, Sandor M, Capel PJA, D eron M, van de Winkel JGJ, Verbeek JS (1996) Impaired IgG-dependent anaphylaxis and Arthus reaction in FcγRIII (CD16) deficient mice. Immunity 5:181-188

Holmdahl R, Andersson M, Goldshmidt TJ, Gustafsson K, Jansson L, Mo JA (1990) Type II collagen autoimmunity in animals and provocations leading to arthritics. Immunol Rev 118:193-232

Holmdahl R, Jansson L, Gullberg D, Rubin K, Forsberg PO, Klareskog L (1985) Incidence of arthritis and autoreactivity of anti-collagen antibodies after immunization of DBA/1 mice with heterologous and autologous collagen II. Clin Exp Immunol 62:639-646

Holmdahl R, Vingsbo C, Malmstr m V, Jansson L, Holmdahl M (1994) Chronicity of arthritis induced with homologous type II collagen (CII) in rats is dependent on anti-CII B-cell activation. J Autoimmun 7:739-752

Kalluri R (1999) Goodpasture syndrome. Kidney Int 55:1120-1122

Kalluri R, Danoff TM, Okada H, Neilson EG (1997) Susceptibility to anti-glomerular basement membrane disease and Goodpasture syndrome is linked to MHC class II genes and the emergence of T-cell-mediated immunity in mice. J Clin Invest 100:2263-2275

Kelly PT, Haponik EF (1994) Goodpasture syndrome: molecular and clinical advances. Medicine 73:171-185

Kleinau S, Martinsson P, Heyman B (2000) Induction and suppression of collagen-induced arthritis is dependent on distinct Fcγ receptors. J Exp Med 191:1611-1616

Lener RA (1987) The role of antiglomerular basement membrane antibody in the pathogenesis of human glomerulonephritis. J Exp Med 126:989-1004

Metzger H (1992) The receptor with high affinity for IgE. Immunol Rev 125:37-48

Miyajima I, Dombrowicz D, Martin TR, Ravetch JV, Kinet JP, Galli SJ (1997) Systemic anaphylaxis in the mouse can be mediated largely through IgG1 and FcγRIII. Assessment of the cardiopulmonary changes, mast cell degranulation, and death associated with active or IgE- or IgG1-dependent passive anaphylaxis. J Clin Invest 99:901-914

Nabozny GH, Baisch JM, Cheng S, Cosgrove D, Griffiths MM, Luthra HS, David CS (1996) HLA-DQ8 transgenic mice are highly susceptible to collagen-induced arthritis: A novel model for human polyarthritis. J Exp Med 183:27-37

Nakamura A, Yuasa T, Ujike A, Ono M, Nukiwa T, Ravetch JV, Takai T (2000) Fcγ receptor IIB-deficient mice develop Goodpasture's syndrome upon immunization with type IV collagen: A novel murine model for autoimmune glomerular basement membrane disease. J Exp Med 191:899-906

Ranges GE, Sriram S, Cooper SM (1985) Prevention of type II collagen-induced arthritis by in vivo treatment with anti-L3T4. J Exp Med 162:1105-1110

Ravetch JV (1994) Fc receptors: Rubor redux. Cell 78:553-560

Ravetch JV, Kinet J-P (1991) Fc receptors. Annu Rev Immunol 9:457-492

Savage CO, Pusey CD, Bowman C, Rees AJ, Lockwood CM (1996) Antiglomerular basement membrane antibody mediated disease in the British Isles. Br Med J 292:1980-1984

Schiller C, Janssen-Graalfs I, Baumann U, Schwerter-Strumpf K, Izui S, Takai T, Schmidt RE, Gessner JE (2000) Mouse FcγRII is a negative regulator of FcγRIII in IgG immune complex triggered inflammation but not in autoantibody induced hemolysis. Eur J Immunol 30:481-490

Seki N, Sudo Y, Yoshioka T, Sugihara S, Fujitsu T, Sakuma S, Ogawa T, Hamaoka T, Senoh H, Fujiwara H (1988) Type II collagen-induced murine arthritis. Induction and perpetuation of arthritis require synergy between humoral and cell-mediated immunity. J Immunol 140:1477-1484

Steblay RW (1979) Anti-glomerular basement membrane glomerulonephritis. Am J Pathol 96:875-878

Stuart JM, Dixon FJ (1983) Serum transfers of collagen-induced arthritis in mice. J Exp Med 158:378-392

Suzuki Y, Shirato I, Okumura K, Ravetch JV, Takai T, Tomino Y, Ra C (1998) Distinct contribution of Fc receptors and angiotensin II-dependent pathways in anti-GBM glomerulonephritis. Kidney Int 54:1166-1174

Svensson L, Jirholt J, Holmdahl R, Jansson L (1998) B cell-deficient mice do not develop type II collagen-induced arthritis (CIA). Clin Exp Immunol 111:521-526

Sylvestre DL, Ravetch JV (1994) Fc receptors initiate the Arthus reaction: Redefining the inflammatory cascade. Science 265:1095-1098

Takai T, Li M, Sylvestre D, Clynes R, Ravetch JV (1994) FcR γ chain deletion results in pleiotrophic effector cell defects. Cell 76:519-529

Takai T, Ono M, Hikida M, Ohmori H, Ravetch JV (1996) Augmented humoral and anaphylactic responses in FcγRII-deficient mice. Nature 379:346–349

Takai T, Ravetch JV (1998) Fc receptor genetics and the manipulation of genes in the study of FcR biology. In: van de Winkel JGJ, Hogarth PM (eds) Immunoglobulin receptors and their physiological and pathological roles in immunity. Kluwer Academic Publishers, Netherland, pp 37-48

Trentham DE, Townes AS, Kang AH (1977) Autoimmunity to type II collagen: an experimental model of arthritis. J Exp Med 146:857-868

Ujike A, Ishikawa Y, Ono M, Yuasa T, Yoshino T, Fukumoto M, Ravetch JV, Takai T (1999) Modulation of IgE-mediated systemic anaphylaxis by low affinity Fc receptors for IgG. J Exp Med 189:1573-1579

van de Winkel JGJ, Capel PJA (1993) Human IgG Fc receptor heterogeneity: Molecular aspects and clinical implications. Immunol Today 14:215-221

Wakayama H, Hasegawa Y, Kawabe T, Hara T, Matsuo S, Mizuno M, Takai T, Kikutani H, Shimokata K (2000) Abolition of anti-glomerular basement membrane antibody-mediated glomerulonephritis in FcRγ-deficient mice. Eur J Immunol 30:1182-1190

Wooley PH, Luthra HS, Stuart JM, David SC (1981) Type II collagen induced arthritis in mice. I. Major histocompatibility complex (I-region) linkage and antibody correlates. J Exp Med 154:688-700

Yuasa T, Kubo S, Yoshino T, Ujike A, Matsumura K, Ono M, Ravetch JV, Takai T (1999) Deletion of FcγRIIB renders H-2^b mice susceptible to collagen-induced arthritis. J Exp Med 189:187-194

25
Receptors involved in human NK cell activation in the process of natural cytotoxicity

Lorenzo Moretta[1,2], Roberto Biassoni[3], Cristina Bottino[3], Maria Cristina Mingari[3,4], and Alessandro Moretta[2]

[1]Istituto Giannina Gaslini, 16148 Genova, Italy
[2]Dipartimento di Medicina Sperimentale, Università degli Studi di Genova, Italy
[3]Istituto Nazionale per la Ricerca sul Cancro, 16132 Genova, Italy
[4]Dipartimento di Oncologia, Biologia e Genetica, Università degli Studi di Genova, Italy

Summary. Natural Killer cells can discriminate between normal cells and cells undergone tumor transformation or viral infection. The discovery of inhibitory receptors specific for major histocompatibility complex class I molecules clarified the molecular basis of this discrimination. The receptors responsible for the 'on' signal in the progress of natural cytotoxicity remained until recently. Several receptors and coreceptors fulfilling this function have now been identified and are illustrated in this contribution.

Key words. Natural Killer cells, activating receptors, natural cytotoxicity

Introduction

During the past several years major advances have been made towards a better understanding of how NK cells function including their ability to discriminate between normal cells and cells undergoing tumor transformation or viral infection (Moretta et al. 1994). These progresses are primarily consequent to the discovery of novel MHC-class I-specific receptors that deliver inhibitory rather than activating signals upon interaction with their MHC ligand. Since MHC class I molecules are expressed on virtually all normal cells, but are frequently missing in tumors and in cells infected with certain viruses, NK cells represent a first, rapid effector mechanism to sense and destroy these dangerous cells. It became also clear that NK cells are regulated by a balance between opposite signals delivered by the MHC-class I-specific inhibitory receptors and by the activating receptors responsible for NK cell triggering. Under normal conditions, the simultaneous engagement of activating and inhibitory receptors results in a dominant inhibitory effect that downregulates signals delivered by the activating pathway (Moretta et al. 1994; Moretta et al. 1996). The signals leading to NK cell inactivation are based on the involvement of tyrosine phosphatases that estinguish the triggering signals dependent on the activity of tyrosine kinases. Ligation of inhibitory receptors leads to tyrosine phosphorilation of immune tyrosine-based inhibitory motifs (ITIM) that recruit tyrosine phosphatases such as the Src homology 2 domain-containing phosphatase (SHP)-1 and SHP-2 which mediate inhibition of various NK cell-mediated effector functions (Long 1999).

In humans, both HLA-specific and non HLA-specific inhibitory receptors have been identified. Those specific for HLA belong to two structurally distinct families of surface molecules: the killer cell Ig-like receptors (KIR) and the killer cell lectin-like receptors (KLR). These receptors are also expressed on a subset of CD8+ CTL with a memory phenotype, in which they may inhibit TCR-mediated T cell activation (Bolland and Ravetch 1999; Mingari et al. 1996; Mingari et al. 1998).

While it became clear that in the absence of inhibition, both NK cell activation and expression of effector function can occur, the receptors responsible for the triggering signals in the process of natural cytotoxicity remained unknown until recently.

In this contribution we will briefly illustrate the recent discovery of a number of activating receptors and coreceptors that allowed to understand the molecular mechanisms involved in natural cytotoxicity.

Natural cytotoxicity receptors

Clues that multiple receptor-ligand interactions may be responsible for NK cell triggering following their interaction with target cells was provided by the heterogeneity of NK cell clones in their ability to lyse different HLA-class I⁻ tumor target cells. The search for these elusive receptors led to the recent identification and molecular characterization of three members of a novel group of receptors collectively referred to as "natural cytotoxicity receptors" (NCR): NKp46, NKp30 and NKp44 (Bottino et al. 2000).

NKp46

NKp46, the prototype of NCR, was discovered by immunizing mice with an NK cell clone displaying strong spontaneous cytotoxicity against different tumor targets. Monoclonal antibodies to NKp46 triggered NK-mediated cytolysis in a redirected killing assay and inhibited to various extents the natural cytotoxicity against different tumor targets (Sivori et al. 1997). Importantly, the NK-mediated lysis of murine target cells was completely abrogated in the presence of anti-

NKp46 mAb, thus suggesting that NKp46 may represent the major receptor allowing human NK cells to recognize ligands on murine cells.

Analysis of cell distribution indicated that NKp46 was expressed by all NK cells (including the CD56^{bright} CD16$^-$ subset) irrespective of their state of activation thus providing an excellent marker for NK cell identification (Sivori et al. 1997).

As revealed by molecular cloning, NKp46 is a novel member of the Ig superfamily, characterized by two C2-type Ig-like domains in the extracellular portion, followed by a stretch of amino acids which connects the ectodomain to the transmembrane region. The transmembrane region contains the positively charged amino acid (Arg) which may be involved in stabilizing the interaction(s) with associated adaptor molecules (Pessino et al. 1998). The cytoplasmic portion does not contain tyrosine-based motifs typically involved in the activation of the signal cascade(s). Biochemical analysis revealed that NKp46 molecules are coupled to the intracytoplasmic transduction machinery by their association with CD3ζ and/or FcϵRIγ adaptor proteins containing immune tyrosine-based activating motifs (ITAM) that become Tyrosine phosphorylated upon receptor engagement (Orloff et al. 1990; Lanier et al. 1989).

The NKp46-encoding gene, similar to all the genes encoding KIR and ILT, maps on human chromosome 19 within the "Leukocyte Receptor Complex" (LRC) at 19q13.4. Murine and rat homologues of human NKp46 have been cloned and found to display a 58% and 59% amino acid identity, respectively, with the human protein (Biassoni et al. 1999; Falco et al. 1999).

NKp30

NKp30 displays several features in common with NKp46. Thus, it is selectively expressed by all NK cells including those expressing the CD56^{bright} CD16$^-$ surface phenotype. Biochemical analysis, revealed a molecular size of approximately 30 kDa thus differing from NKp46. In addition, mAb-mediated surface modulation of NKp46 did not affect the expression of NKp30. Similar to NKp46 and CD16, NKp30 associated with CD3ζ chains (Pende et al. 1999).

Cloning of the cDNA encoding NKp30 revealed a novel member of the Ig superfamily. The extracellular portion is characterized by a single domain of the V-type and by a region rich in hydrophobic amino acids that connects the IgV-like domain with the transmembrane portion. The transmembrane region contains the positively charged amino acid Arg, probably involved in the association with CD3ζ chains, whereas the cytoplasmic portion lacks typical ITAM consensus sequences. The NKp30-encoding gene maps on chromosome 6 in the TNF cluster of the MHC gene complex (Pende et al. 1999).

NKp44

NKp44 represents a third member of the NCR family displaying a molecular size of approximately 44 kDa. Also NKp44 could induce triggering of NK mediated cytotoxicity upon cross-linking by specific mAbs (Vitale et al. 1998). However, the expression of NKp44 is restricted to activated NK cells cultured in the presence of IL2 while it is absent in fresh peripheral blood NK cells. Therefore, NKp44 represents the first marker specific for activated human NK cells (Vitale et al. 1998). In addition, following culture in the presence of IL2, NK cells acquire increased cytolytic activity against NK susceptible targets. This may reflect, at least in part, the de novo expression of triggering receptors such as NKp44 that allow NK cells to recognize additional ligands on target cells. Molecular cloning of NKp44 revealed a novel member of the Ig-SF characterized by a single extracellular V-type domain (Cantoni et al. 1999). The membrane-proximal portion of NKp44 contains a high proportion of Pro, Ser and Thr that may confer an extended open conformation. The gene encoding NKp44, similar to that coding for NKp30, is located on human chromosome 6. NKp44 does not contain ITAM in the cytoplasmic portion while it is characterized by a transmembrane region containing a charged amino acid (Lys) possibly involved in the association with KARAP/DAP12 signal transducing adapter proteins.

Involvement of NCR in tumor cell lysis

NK cell clones frequently display heterogeneity in their ability to kill a given HLA-class I-negative target cell. In some individuals, two distinct groups of NK cell clones could be identified on the basis of the ability to kill a given HLA-class I⁻ NK susceptible target cell. Thus, some clones displayed strong cytolytic activity while other were poorly cytolytic. The surface density of NKp46 was found to be markedly different in the two groups of clones. Thus, clones displaying strong cytolytic activity expressed high levels of NKp46 whereas the poorly cytolytic ones were characterized by low NKp46 surface density. Importantly, NKp46^{bright} and NKp46^{dull} clones were also characterized by a parallel high and low level of expression of NKp30 and NKp44 receptors (Sivori et al. 1999). Remarkably, The NCR surface density did not correlate with the actual cytolytic potential of NK cells. Thus, cytolytic responses induced by anti-CD16 mAbs was comparable in NKp46^{bright} and NCR^{dull} NK clones.

A correlation between the NCR^{bright} phenotype and strong natural cytotoxicity could be consistently detected when NK cells were tested against most target cells (including melanomas, EBV-LCL, lung carcinomas, PHA blasts and xenogeneic target cells). However, exceptions to this rule exist. Target cells including certain T cell lymphomas (e.g. Jurkat and H9 cells) and epithelial tumor cell lines were efficiently killed by both NCR^{bright} and NCR^{dull} NK clones. It appeared conceivable that recognition and killing of these target cells could depend not only on NCR but also on additional triggering receptors expressed by NCR[dull] clones. These data suggested that the analysis of NCR[dull] clones may be crucial to identify novel triggering receptors (Sivori et al. 1999) (see below).

Experiments using mAbs to different NCRs revelead that they play a major role as receptors involved in recognition and lysis of the majority of tumor cells analyzed. In most cases, blocking of individual NCR resulted in partial inhibition of cytotoxicity, while strong inhibition required the simultaneous mAb-mediated masking of different NCR. This indicates that different NCR cooperate in target cell recognition and killing. The extent of such cooperation appears to depend upon the type and/or the density of the ligands expressed on target cells. Although the NCR ligands have not been molecularly identified so far, the cooperative effect of different NCR implies that most target cells express multiple ligands for NCR. An exception is represented by xenogeneic (murine) target cells. Thus, lysis of mouse targets by human NK cells could be blocked by mAb-mediated masking of NKp46 only (Pessino et al. 1998). The finding that human NK cells recognize ligands on xenogeneic cells, together with the identification of a mouse and rat homologue of NKp46 suggested that the NKp46/ligand interaction is conserved during evolution. In this respect, the occurrence of triggering interactions involving human NK cells may represent a serious problem in xenotransplantation. For example human NK cells are able to kill normal (MHC-class I+) porcine targets due to the failure of pig MHC class I molecules to interact with human HLA-class I-specific inhibitory receptors.

NKG2-D

NKG2-D is a member of the NKG2 family encoded within the "NK complex" on human chromosome 12. NKG2A, C, E and F show a high degree of identity each other and form heterodimers with CD94, some of which have been shown to function as receptor for HLA-E (Carretero et al. 1997; Lazetic et al. 1996; Cantoni et al. 1998; Braud et al. 1998). On the contrary, NKG2D is only distantly related to the other members of the family and does not couple with CD94 being expressed as a homodimer. NKG2D is associated with a newly identified adaptor protein termed DAP10 (Wu et al. 1999; Chang et al. 1999). Different from NCR that are confined to NK cells, NKG2D is also expressed by virtually all TCRγ/δ[+] and CD8[+] TCRα/β[+] cells. NKG2D mediates potent NK cell triggering in redirected killing assays. Blocking of NKG2D by specific mAbs leads to inhibition of the NK cell-mediated cytotoxicity against certain target cells. The target cell ligands for NKG2D have been identified. They are represented by the closely related molecules MICA and MICB that are encoded within the human MHC. MICA/B are stress-inducible molecules primarily expressed in epithelial tumors (Bauer et al. 1999; Groh et al. 1999). The interaction between NKG2D and MICA/B leads to NK cell activation while mAb-mediated disruption of this interaction inhibits NK-mediated cytotoxicity. The actual role of NKG2D in the

NK-mediated natural cytotoxicity has recently been evaluated in human NK cell clones and compared to that of NCR (Pende et al. in press). These studies indicate that in NCR^{bright} clones killing of various tumors was mostly NCR-dependent while killing of other targets required triggering signals via both NCR and NKG2D. Further analysis of NCR^{dull} clones indicated that both the expression and the function of NKG2D did not correlate with the surface density of NCR (Pende et al. in press). Important information of the function of NKG2D has been obtained by the study of NCR^{dull} clones. In spite of the low expression of NCR, these clones could efficiently kill certain tumors. For example, the cytolytic activity of NCR^{dull} clones against MICA$^+$ target cells such as HeLa and IGROV was virtually abrogated by anti-NKG2D mAb. Thus suggesting that lysis of these tumors by NCR^{dull} clones, is mostly NKG2D-dependent. Thus, the function of NKGD2 is complementary to that of NCR.

2B4

2B4 was first described as a surface molecule recognized by the monoclonal antibodies PP35 (Moretta et al. 1992) and C1.7 (Valiante and Trinchieri 1993). These mAbs immunoprecipitated a 70 kD glycoprotein expressed by all NK cells and by a CD8+ T cell subset. mAb-mediated cross-linking could strongly enhance the NK-mediated cytotoxicity while it had no effect on CD8+ T cells. Molecular cloning revealed that the 2B4-encoding gene is located on chromosome 1 in the CD2 subfamily cluster (Kubin et al. 1999). It is characterized by one membrane-distal IgV type domain and one membrane proximal Ig-C2 type domain. In agreement with the lack of association of 2B4 with ITAM-containing polypeptides, the transmembrane portion does not contain charged amino acids. 2B4 is characterized by a long cytoplasmic tail containing four tyrosine-based motifs (TxYxxI/V). CD48 has been identified as the putative natural ligand of 2B4 (Nakajima et al. 1999).

In human NK cells, 2B4 displays a remarkable functional heterogeneity. For example, the ability of anti-2B4 mAb to induce NK cell activation in redirected killing assays was confined to a fraction of NK cell clones. These results have recently been clarified: the apparent functional heterogeneity of 2B4 molecules is actually reflecting clonal differences in the co-engagement of triggering receptors. Only NCR^{bright} clones could be triggered by anti-2B4 mAb in a redirected killing assay (Sivori et al. 2000). No triggering of NCR^{dull} clones occurred, although they expressed comparable surface densities of 2B4. In addition, mAb-mediated modulation of NKp46 did not affect the surface expression of 2B4, but resulted in NK cell unresponsiveness to anti-2B4 mAb (Sivori et al. 2000). These data support the notion that in human NK cell-mediated natural cytotoxicity, NCRs function as triggering receptors while 2B4 acts as a co-receptor.

Controversial information exists on the molecular mechanisms involved in the 2B4-mediated signaling. Thus, by the use of cell transfectants, 2B4 has been shown to associate with the SH2D1A protein also called SLAM-associated protein (SAP), and with the SHP-2 phosphatase (Tangye et al. 1999). A recent study using normal NK cells confirmed that tyrosine-phosphorylation of 2B4 leads to association with SAP. However, different from cell transfectants, 2B4 associated with SHP-1 rather than with SHP-2 (Sivori et al. 2000). Taken together, these data suggest that, in normal NK cells, SAP may sustain the 2B4-mediated triggering responses by preventing the generation of inhibitory signals mediated by SHP-1. In normal human NK cells 2B4 constitutively associates with the Linker for Activation of T cells (LAT) (Bottino et al. 2000). Specific engagement of 2B4, following cell triggering with an anti-2B4 mAb, resulted in tyrosine phosphorylation of both 2B4 and of the associated LAT molecules. Moreover, tyrosine phosphorylation of LAT led to the recruitment of intracytoplasmic signaling molecules including PLCγ and Grb2.

An altered function of 2B4 has been shown to represent a major defect in the X-linked lymphoproliferative disease (XLP). XLP is a severe inherited immune deficiency, characterized by abnormal immune responses to the Epstein Barr virus (EBV). After exposure to EBV, XLP males frequently (80%) succumb to fulminant infectious mononucleosis. The gene defective in XLP has recently been identified: it is located at Xq25 and, in normal individuals, encodes for the SH2D1A/ SAP protein (see above). SAP is expressed in T and NK lymphocytes. Different mutations of the SH2D1A gene have been identified the result of which is either the absence of SAP molecule or the presence of truncated, non-functional products (Bottino et al. 2000; Nichols et al. 1998; Coffey et al. 1998). It has been proposed that the SH2D1A/SAP protein functions as a regulator of the signal transduction pathways initiated by the signaling lymphocytic activation molecule (SLAM) and by 2B4. This suggested that, in XLP patients, 2B4 molecules could be non-functional owing to the lack of SAP (Sayos et al. 1998). Indeed, the signaling pathway initiated by 2B4 was dramatically altered. But there was not simply a "lack of function" as the engagement of 2B4 (by specific mAbs or its ligand CD48) resulted in the generation of inhibitory rather than activating signals (Parolini et al. 2000). These signals blocked not only the spontaneous NK cytotoxicity, but also the target cell lysis induced via CD16 or NCR. As a consequence, XLP NK cells could efficiently kill CD48⁻ but not CD48⁺ target cells.

It is of note that high densities of CD48 are expressed on EBV-infected LCL. Indeed, XLP-NK cells are unable to kill EBV⁺ B cell lines. The cytolytic activity against HLA class I⁻ LCL could be completely restored by mAb-mediated disruption of the 2B4/CD48 interactions (Parolini et al. 2000). In XLP-NK cells 2B4 did not associate with SAP whereas, similarly, to 2B4 molecules isolated from normal NK cells, it did associate with SHP-1 phosphatase. Thus, leading to downregulation of different activation pathways including those mediated by NCR or CD16. Preliminary data also indicate that 2B4 may also negatively regulate T cell-mediated responses. Therefore, it is conceivable that specific CTL responses against EBV⁺ B cells may be impaired. Thus, the altered function of 2B4 in XLP patients may account for a general inability of different cytolytic effector cells to control EBV infections.

Conclusion

In conclusion, the identification and the functional characterization of various surface molecules regulating the NK cell function may help us to understand the NK cell physiology and may also shed light on the pathogenesis of certain immune disorders. Perhaps more importantly, it may offer also novel tools for therapeutic interventions in these diseases.

Acknowledgment

This work was supported by grants awarded by Associazione Italiana per la Ricerca sul Cancro (A.I.R.C.), Istituto Superiore di Sanità (I.S.S.), Ministero della Sanità, and Ministero dell'Università e della Ricerca Scientifica e Tecnologica (M.U.R.S.T.), MURST 5%–CNR Biotechnology program 95/95 and Consiglio Nazionale delle Ricerche, Progetto Finalizzato Biotecnologie. Also the financial support of Telethon-Italy (grant no.E.0892) is gratefully acknowledged.

References

Bauer S, Groh V, Wu J, Steinle A, Phillips JH, Lanier LL, Spies T (1999) Activation of NK cells and T cells by NKG2D, a receptor for stress-inducible MICA. Science 285:727-729

Biassoni R, Pessino A, Bottino C, Pende D, Moretta L, Moretta A (1999) The murine homologue of the human NKp46, a triggering receptor involved in the induction of natural cytotoxicity. Eur J Immunol 29:1014-1020

Bolland S, Ravetch JV (1999) Inhibitory pathways triggered by ITIM-containing receptors. Adv Immunol 72:149-177

Bottino C, Augugliaro R, Castriconi R, Nanni M, Biassoni R, Moretta L, Moretta A (2000) Analysis of the molecular mechanism involved in 2B4-mediated NK cell activation: evidence that human 2B4 is physically and functionally associated with the linker for activation of T cells (LAT). Eur J Immunol 30:3718-3722

Bottino C, Biassoni R, Millo R, Moretta L, Moretta A (2000) The human Natural Cytotoxicity Receptors (NCR) that induce HLA Class I-independent NK cell triggering. Human Immunol 61:1-6

Braud VM, Allan DSJ, O'Callaghan CA, Soderstrom K, D'Andrea A, Ogg GS, Lazetic, S, Young, NT, Bell JI, Phillips JH, Lanier, LL, McMichael AJ (1998) HLA-E binds to natural killer cell receptors CD94/NKG2A, B and C. Nature 391:795-799

Cantoni C, Biassoni R, Pende D, Sivori S, Accame L, Pareti L, Semenzato G, Moretta L, Moretta A, Bottino C (1998) The activating form of CD94 receptor complex. CD94 covalently associates with the Kp39 protein that represents the product of the NKG2-C gene. Eur J Immunol 28:327-338

Cantoni C, Bottino C, Vitale M, Pessino A, Augugliaro R, Malaspina A, Parolini S, Moretta L, Moretta A, Biassoni R (1999) NKp44, a triggering receptor involved in tumor cell lysis by activated human Natural Killer cells, is a novel member of the immunoglobulin superfamily. J Exp Med 189:787-796

Carretero M, Cantoni C, Bellón T, Bottino C, Biassoni R, Rodríguez A, Pérez-Villar JJ, Moretta L, Moretta A, López-Botet M (1997) The CD94 and NKG2-A C-type lectins covalently assamble to form a a NK cell inhibitory receptor for HLA class I molecules. Eur J Immunol 27:563-567

Chang C, Dietrich J, Harpur AG, Lindquist JA, Haude A, Loke YW, King A, Colonna M, Trowsdale J, Wilson MJ (1999) KAP10, a novel transmembrane adapter protein genetically linked to DAP12 but with unique signaling properties. J Immunol 163:4651-4654

Coffey AJ, Brooksbank RA, Brandau O, Oohashi T, Howell GR, Bye JM, Cahn AP, Durham J, Heath P, Wray P, Pavitt R, Wilkinson J, Leversha M, Huckle E, Shaw-Smith CJ, Dunham A, Rhodes S, Schuster V, Porta G, Yin L, Serafini P, Sylla B, Zollo M, Franco B, Bolino A, Seri M, Lanyi A, Davis JR, Webster DW, Harris A, Lenoir G, de St Basile G, Jones A, Behloradsky BH, Achatz H, Murken HJ, Fassler R, Sumegi J, Romeo G, Vaudin M, Ross MT, Meindl A, Bentley DR (1998) Host response to EBV infection in X-linked lymphoproliferative disease results from mutations in an SH2-domain encoding gene. Nat Genet 20:129-135

Falco M, Cantoni C, Bottino C, Moretta A, Biassoni R (1999) Identification of the rat homologue of the human NKp46 triggering receptor. Immunol Lett 68:411-414

Groh V, Rhinehart R, Secrist H, Bauer S, Grabstein KH, Spies T (1999) Broad tumor-associated expression and recognition by tumor-derived $\gamma\delta$ T cells of MICA and MICB. Proc Natl Acad Sci USA 96:6879-6884

Houchins JP, Yabe T, McSherry C, Bach FH (1991) DNA sequence analysis of NKG2, a family of related cDNA clones encoding type II integral membrane proteins on human natural killer cells. J Exp Med 173:1017-1020

Kubin MZ, Parsley DL, Din W, Waugh JY, Davis-Smith T, Smith CA, Macduff BM, Armitage RJ, Chin W, Cassiano L, Borges L, Petersen M, Trinchieri G, Goodwing RG (1999) Molecular cloning and biological characterization of NK cell activation-inducing ligand, a counterstructure for CD48. Eur J Immunol 29:3466-3477

Lanier LL, Yu G, Phillips JH (1989) Co-association of CD3 ζ with a receptor (CD16) for IgG Fc on human natural killer cells. Nature 342:803-805

Lazetic S, Chang C, Houchins JP, Lanier LL, Phillips JH (1996) Human Natural killer cell receptors involved in MHC class I recognition are disulphide-linked heterodimers of CD94 and NKG2 subunits. J Immunol 157:4741-4745

Long EO (1999) Regulation of immune responses through inhibitory receptors. Annu Rev Immunol 17:875-904

Mingari MC, Moretta A, Moretta L (1998) Regulation of KIR expression in human T lymphocytes. A safety mechanism which may impair protective T cell responses. Immunol Today 19:153-157

Mingari MC, Schiavetti F, Ponte M, Vitale C, Maggi E, Romagnani S, Demarest J, Pantaleo G, Fauci AS, Moretta L (1996) Human CD8+ T lymphocyte subsets that express HLA class I-specific inhibitory receptors represent oligoclonally or monoclonally expanded cell populations. Proc Natl Acad Sci USA 93:12433-12438

Moretta A, Bottino C, Tripodi G, Vitale M, Pende D, Morelli L, Augugliaro R, Barbaresi M, Ciccone E, Millo R, Moretta L (1992) Novel surface molecules involved in human NK cell activation and triggering of the lytic machinery. Int J Cancer Suppl 7:6-10

Moretta A, Bottino C, Vitale M, Pende D, Biassoni R, Mingari MC, Moretta L (1996) Receptors for HLA-class I molecules in human Natural Killer cells. Annu Rev Immunol 14:619-648

Moretta L, Ciccone E, Mingari MC, Biassoni R, Moretta A (1994) Human NK cells: origin, clonality, specificity and receptors. Adv Immunol 55:341-380

Nakajima H, Cella M, Langen H, Friedlein A, Colonna M (1999) Activating interactions in human NK cell recognition: the role of 2B4-CD48. Eur J Immunol 29:1676-1683

Nichols KE, Harkin DP, Levitz S, Krainer M, Kolquist KA, Genovese C, Bernard A, Ferguson M, Zuo L, Snyder E, Buckler AJ, Wise C, Ashley J, Lovett M, Valentine MB, Look AT, Gerald W, Housman DE, Haber DA (1998) Inactivating mutations in an SH2 domain-encoding gene in X-linked lymphoproliferative syndrome. Proc Natl Acad Sci USA 95:13765-13770

Orloff DG, Ra CS, Frank SJ, Klausner RD, Kinet JP (1990) Family of disulphide-linked dimers containing the ζ and η chains of the T-cell receptor and the γ chain of Fc receptors. Nature 347:189-191

Parolini S, Bottino C, Falco M, Augugliaro R, Silvia Giliani S, Franceschini R, Ochs HD, Wolf H, Bonnefoy J-Y, Biassoni R, Moretta L, Notarangelo LD, Moretta A (2000) X-linked lymphoproliferative disease: 2B4 molecules displaying inhibitory rather than activating function are responsible for the inability of NK cells to kill EBV-infected cells. J Exp Med 192:337-346

Pende D, Cantoni C, Rivera P, Vitale M, Castriconi R, Marcenaro S, Nanni M, Biassoni R, Bottino C, Moretta A, Moretta L Role of NKG2D in tumor cell lysis mediated by human NK cells: cooperation with natural cytotoxicity receptors and capability of recognizing tumors of non epithelial origin. Eur J Immunol (in press)

Pende D, Parolini S, Pessino A, Sivori S, Augugliaro R, Morelli L, Marcenaro E, Accame L, Malaspina A, Biassoni R, Bottino C, Moretta L, Moretta A (1999) Identification and molecular characterization of NKp30, a novel triggering receptor involved in natural cytotoxicity mediated by human natural killer cells. J Exp Med 190:1505-1516

Pessino A, Sivori S, Bottino C, Malaspina A, Morelli L, Moretta L, Biassoni R, Moretta A (1998) Molecular cloning of NKp46: a novel member of the immunoglobulin superfamily involved in triggering of natural cytotoxicity. J Exp Med 188:953-960

Sayos J, Wu C, Morra M, Wang N, Zhang X, Allen D, van Schaik S, Notarangelo L, Geha R, Roncarolo MG, Oettgen H, De Vries JE, Aversa G, Terhorst C (1998) The X-linked lymphoproliferative-disease gene product SAP regulates signals induced through the co-receptor SLAM. Nature 395:462-469

Sivori S, Parolini S, Falco M, Marcenaro E, Biassoni R, Bottino C, Moretta L, Moretta A (2000) 2B4 functions as a co-receptor in human natural killer cell activation. Eur J Immunol 30:787-793

Sivori S, Pende D, Bottino C, Marcenaro E, Pessino A, Biassoni R, Moretta L, Moretta A (1999) NKp46 is the major triggering receptor involved in the natural cytotoxicity of fresh or cultured human natural killer cells. Correlation between surface density of NKp46 and natural cytotoxicity against autologous, allogeneic or xenogeneic target cells. Eur J Immunol 29:1656-1666

Sivori S, Vitale M, Morelli L, Sanseverino L, Augugliaro R, Bottino C, Moretta L, Moretta A (1997) p46, a novel Natural Killer cell-specific surface molecule which mediates cell activation. J Exp Med 186:1129-1136

Tangye SG, Lazetic S, Woollatt E, Sutherland GR, Lanier LL, Phillips JH (1999) Human 2B4, an activating NK cell receptor, recruits the protein tyrosine phosphatase SHP-2 and the adaptor signaling protein SAP. J Immunol 162:6981-6985

Valiante NM, Trinchieri G (1993) Identification of a novel signal transduction surface molecule on human cytotoxic lymphocytes. J Exp Med 178:1397-1406

Vitale M, Bottino C, Sivori S, Sanseverino L, Castriconi R, Marcenaro R, Augugliaro R, Moretta L, Moretta A (1998) NKp44, a novel triggering surface molecule specifically expressed by activated natural killer cells is involved in non-MHC restricted tumor cell lysis. J Exp Med 187:2065-2072

Wu J, Song Y, Bakker AB, Bauer S, Spies T, Lanier LL, Phillips JH (1999) An activating immunoreceptor complex formed by NKG2D and DAP10. Science 285:730-732

26
The regulation of PD-1/PD-L1 pathway and autoimmune diseases

Taku Okazaki[1], Yoshiko Iwai[1], Hiroyuki Nishimura[1], and Tasuku Honjo[1]

[1]Department of Medical Chemistry, Graduate school of Medicine, Kyoto University, Yoshida-Konoe, Sakyo-ku, Kyoto, 606-8501, Japan

Summary. PD-1 is an immuno-inhibitory receptor which belongs to the immunoglobulin superfamily and expressed on activated T, B and myeloid cells. Engagement of the PD-1 receptor with its membrane bound ligand (PD-L1) of the B7 family inhibits the proliferation of anti-CD3 stimulated T cells as well as anti-IgM stimulated B cells (Freeman et al. 2000 and our unpublished observation). Disruption of PD-1 gene in C57BL/6 mice caused typical lupus-like glomerulonephritis and destructive arthritis as they age (Nishimura et al. 1999) while in BALB/c mice caused autoantibody mediated dilated cardiomyopathy with severely impaired contraction and sudden death by congestive heart failure. Affected hearts showed diffuse deposition of IgG on the surface of cardiomyocytes. All of the affected PD-1$^{-/-}$ mice exhibited high-titered circulating IgG autoantibodies reactive to a 33-kDa protein expressed specifically on the surface of cardiomyocytes (Nishimura et al. 2001). These results indicate that PD-1 may be an important factor contributing to the prevention of autoimmune diseases.

Key words. Autoimmune disease, Dilated cardiomyopathy, Autoantibody

Disruption of PD-1 gene causes dilated cardiomyopathy

Dilated cardiomyopathy is a chronic disorder of the heart muscle characterized by a poorly contractile and dilated left venticle. The diagnosis is based primarily on clinical criteria and on the exclusion of identifiable underlying causes (Richardson et al. 1996). Consequently, patients with dilated cardiomyopathy represent a heterogeneous group affected to various degrees by genetic, viral, immunological, and environmental factors (Matsumori 1997), which has complicated the distinction of underlying pathogenic mechanisms. In recent years, evidence has accumulated in support of an immune mechanism in a certain fraction of the

212

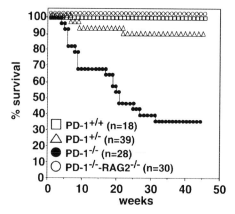

Fig. 1. Survival curves of BALB/c-PD-1$^{-/-}$ mice.

patients, whose sera are shown to contain antibodies against self-antigens (Caforio et al. 1994; Klein et al. 1984; Magnusson et al. 1990; Schulze et al. 1989). On the BALB/c background, PD-1$^{-/-}$ mice started to die as early as 5 weeks of age. By 30 weeks, two thirds of PD-1$^{-/-}$ and 10% of PD-1$^{+/-}$ mice had died, while all of the PD-1$^{+/+}$ controls survived (Fig. 1). In contrast, this early death has not been observed in either RAG2$^{-/-}$-PD-1$^{-/-}$ or diseased C57BL/6-PD-1$^{-/-}$ mice. Autopsy examination revealed that all diseased mice exhibited massively enlarged hearts and various degree of hepatomegaly, suggesting that the cause of death was congestive heart failure. Histological examination revealed the right ventricular walls of PD-1$^{-/-}$ mice were thinner than those of the control mice, and both ventricles were dilated about two-fold in diameter. Echocardiographic studies revealed that the pump function of the dilated hearts of PD-1$^{-/-}$ mice was severely impaired in a manner that conformed to a diagnosis of dilated cardiomyopathy (Nishimuraet al. 2001).

Deposition of IgG on the surface of the cardiomyocytes and existence of circulating autoantibodies reactive to 33kDa heart protein

Since all of the PD-1$^{-/-}$ mice on the BALB/c-RAG2$^{-/-}$ background remained complete health, development of heart disease was attributed to the function of T and/or B lymphocytes. Immunofluorescent analysis revealed the linear deposition of IgG, but little IgM, together with C3 complement surrounding the cardiac myocytes in the affected hearts. The linear, as oposed to the granular, staining pattern of IgG implied the binding of tissue-specific autoantibodies rather than the deposition of immune complex. Indeed, there was little detectable IgG deposition in other organs including renal glomeruli.

To confirm the presence of a specific autoimmune reaction against the heart, we examined mice for the presence of autoantibodies against the heart tissue. Sera from all of the sick PD-1$^{-/-}$ mice exhibited high-titered IgG reactive to a 33kDa protein in the normal heart extract. In contrast, the autoantibodies were detected in none of the sera from age-matched BALB/c-PD-1$^{+/+}$, healthy BALB/c-PD-1$^{-/-}$, and C57BL/6-PD-1$^{-/-}$ mice (Fig. 2A). The 33kDa autoantigen appears to be specific for the heart tissue, since it was not detected in other tissue such as liver, kidney or skeletal muscle by the same sera from sick BALB/c-PD-1$^{-/-}$ mice. No other common autoantibodies including anti-dsDNA could be detected in the sera of PD-1$^{-/-}$ mice. To characterize properties of the 33kDa antigen recognized by autoantibodies, we purified the antibodies bound to the 33kDa protein and used them for staining cardiac myocytes. As shown in Fig. 2B, autoantibodies specific to the 33kDa protein stained the surface of cardiac myocytes with dotted structure which, in a way, resembles the staining pattern with wheat germ known to bind to the transverse (T) tube. Although we cannot conclude the exact property of the 33kDa antigen, it is likely to be localized with some structure or the surface of cardiac myocytes, suggesting its involvement in cardiac function. Blocking such functional molecules by antibody binding would cause dysfunction of cardiac muscle.

At present, it remains to be verified whether the autoantibody specific for the heart has the primary role for the disease or whether autoreactive effector T cells

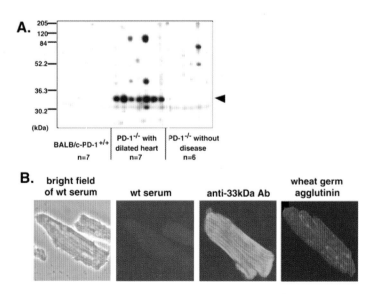

Fig. 2A,B. Circulating autoantibodies against heart in sick PD-1$^{-/-}$ mice. A: Sera from BALB/c-PD-1$^{-/-}$ with dilated heart recognize 33kDa heart protein by western blotting. B: Anti-33kDa heart protein antibody specifically stains nonpermiabilized isolated cardiomyocytes.

are additionally involved in the cardiac tissue damage. Nevertheless, the present results provide the first indication that the typical dilated cardiomyopathy can develop by an autoimmune mechanism due to the endogenous dysregulation of the immune system. The dilated cardiomyopathy is progressive and life-threatening, and no effective therapy is available at present with the exception for heart transplantation. Present results raise the possibility that some forms of cardimyopathy may have an autoimmune basis. The identification of possible autoantigen(s) may open new therapeutic approaches for this significant disease.

References

Caforio AL, Keeling PJ, Zachara E, Mestroni L, Camerini F, Mann JM, Bottazzo GF, McKenna WJ (1994) Evidence from family studies for autoimmunity in dilated cardiomyopathy. Lancet 344: 773-777

Freeman GJ, Long AJ, Iwai Y, Bourque K, Chernova T, Nishimura H, Fitz L J, Malenkovich N, Okazaki T, Byrne MC, Horton HF, Fouser L, Carter L, Ling V, Bowman MR, Carreno BM, Collins M, Wood CR, Honjo T (2000) Engagement of the PD-1 immunoinhibitory receptor by a novel B7 family member leads to negative regulation of lymphocyte activation [In Process Citation]. J Exp Med 192: 1027-1034

Klein R, Maisch B, Kochsiek K, Berg PA (1984) Demonstration of organ specific antibodies against heart mitochondria (anti-M7) in sera from patients with some forms of heart diseases. Clin Exp Immunol 58: 283-292

Magnusson Y, Marullo S, Hoyer S, Waagstein F, Andersson B, Vahlne A, Guillet JG, Strosberg AD, Hjalmarson A, Hoebeke J (1990) Mapping of a functional autoimmune epitope on the beta 1-adrenergic receptor in patients with idiopathic dilated cardiomyopathy. J Clin Invest 86: 1658-1663

Matsumori A (1997) Molecular and immune mechanisms in the pathogenesis of cardiomyopathy--role of viruses, cytokines, and nitric oxide. Jpn Circ J 61: 275-291

Nishimura H, Nose M, Hiai H, Minato N, Honjo T (1999) Development of lupus-like autoimmune diseases by disruption of the PD-1 gene encoding an ITIM motif-carrying immunoreceptor. Immunity 11: 141-151

Nishimura H, Okazaki T, Tanaka Y, Nakatani K, Hara M, Matsumori A, Sasayama S, Mizoguchi A, Hiai H, Minato N, Honjo T (2001) Autoimmune dilated cardiomyopathy in PD-1 receptor deficient mice. Science 291:319-322

Richardson P, McKenna W, Bristow M, Maisch B, Mautner B, O'Connell J, Olsen E, Thiene G, Goodwin J, Gyarfas I, Martin I, Nordet P (1996) Report of the 1995 World Health Organization/International Society and Federation of Cardiology Task Force on the Definition and Classification of cardiomyopathies. Circulation 93: 841-842

Schulze K, Becker BF, Schultheiss HP (1989) Antibodies to the ADP/ATP carrier, an autoantigen in myocarditis and dilated cardiomyopathy, penetrate into myocardial cells and disturb energy metabolism in vivo. Circ Res 64: 179-192

27

Identification of the functional recognition site on MHC class I for a NK cell lectin-like receptor

Naoki Matsumoto[1], Motoaki Mitsuki[1], Kyoko Tajima[1], Wayne M. Yokoyama[2], and Kazuo Yamamoto[1]

[1]Department of Integrated Biosciences, Graduate School of Frontier Sciences, The University of Tokyo, 7-3-1 Hongo, Bunkyo-ku, Tokyo 113-0033, Japan
[2]Howard Hughes Medical Institute, Department of Internal Medicine, Washington University School of Medicine, 660 S. Euclid, St. Louis, MO 63110, USA

Summary. NK cells monitor expression of MHC class I using two structurally distinct groups of receptors and discriminate normal from infected or transformed cells. Recent structural and biochemical studies have revealed the mode of ligand recognition by the two groups of receptors. Mouse Ly49A, a prototypic member of the C-type lectin-like receptors, binds a concave region of its MHC class I ligand H-2D^d partially overlapping with the CD8-binding site. KIR2DL2, a human immunoglobulin-like NK cell receptor, binds top of its MHC class I ligand HLA-Cw3 which overlaps with the binding site for the T-cell receptor. This paper focuses on recent progress in understanding the structural basis of ligand recognition by Ly49A and discusses the similarities and differences between the two groups of NK cell receptors in the recognition of their MHC class I ligands.

Key words.. β2-Microglobulin, Inhibitory receptors, Carbohydrate, Polymorphism

Introduction

NK cells express receptors for MHC class I molecules to monitor their expression on target cells (Yokoyama, 1999). When NK cells encounter a target that has lost normal expression of MHC class I, the NK cells are released from inhibition to kill the target. The MHC class I receptors on NK cells are classified in two structurally distinct groups (Lanier, 1998). One group that belongs to immunoglobulin (Ig) superfamily is prominent in human and contains human killer cell immunoglobulin-like receptors (KIR). The other group that belongs to C-type lectin superfamily is found in human and rodents. The latter group includes Ly49 family in rodents and CD94/NKG2 in human and rodents. Recent structural and biochemical studies have revealed the molecular basis for recognition of MHC class I by the two structurally distinct NK cell receptors. This contribution reviews our most recent work identifying the functional site for the C-type lectin-like receptor Ly49A on its MHC class I ligand and discusses similarities and differences in the recognition of MHC class I by two classes of NK cell receptors.

Ly49A recognition of a conformational epitope in $\alpha 1/\alpha 2$ domain of MHC class I

Mouse NK cells express Ly49 family receptors, which have an extracellular domain homologous to C-type lectin, to recognize classical MHC class I. The Ly49 family consists of more than 10 members that are homodimers of type II transmembrane proteins (Smith et al. 1994; Wong et al. 1991). Ly49A, the prototype member of this family, is an inhibitory receptor specific for the mouse MHC class I molecules H-2D^d and H-2D^k (Karlhofer et al. 1992). The expression of these MHC class I molecules on the target inhibits its killing by NK cells that express Ly49A.

MHC class I consists of three structural domains: $\alpha 1/\alpha 2$ domain, which has a peptide-binding cleft, $\alpha 3$ domain and $\beta 2$-microglobulin ($\beta 2$m) (Bjorkman et al. 1987). Several lines of evidence suggest the involvement of the $\alpha 1/\alpha 2$ in the Ly49A recognition of MHC class I. The 34-5-8S monoclonal antibody (mAb), which recognizes $\alpha 1/\alpha 2$ domain of H-2D^d, inhibits the recognition while 34-2-12S mAb against the $\alpha 3$ domain does not (Karlhofer et al. 1992). The natural recombinant MHC class I molecule dm-1, which has $\alpha 1$ and NH$_2$-terminal half of $\alpha 2$ domain from H-2D^d and rest of the regions from non-Ly49A ligand H-2L^d, functionally interacts with Ly49A (Karlhofer et al. 1994). Ly49A only recognizes H-2D^d in a complex with peptides which bind peptide-binding cleft in $\alpha 1/\alpha 2$ domain (Correa and Raulet 1995; Orihuela et al. 1996). A panel of artificial recombinant MHC class I between H-2D^d and non-Ly49A ligand K^d has localized polymorphic determinant in β-sheet region in $\alpha 1/\alpha 2$ domain (Matsumoto et al. 1998). These data suggest a pivotal role of $\alpha 1/\alpha 2$ domain in the interaction.

Critical role of the $\beta 2$-microglobulin subunit in Ly49A recognition of H-2D^d

Besides the established role of heavy chain, especially of $\alpha 1/\alpha 2$ domain in Ly49A recognition of the ligand MHC class I, the role of the other subunit, $\beta 2$m, has been uncertain except for the essential role of $\beta 2$m in cell surface expression of MHC class I. To test for the direct involvement of $\beta 2$m in the Ly49A-H-2D^d interaction, we have examined the effect of an anti-$\beta 2$m mAb in the interaction. S19.8 mAb recognizes an allelic form of β2m found in C57BL strains of mice, C57BL/6 and C57BL/10 (Margulies et al. 1983). Fortunately our experimental system includes the C1498 lymphoma derived from a C57BL/6 mouse that was transfected with H-2D^d. Quite unexpectedly, S19.8 mAb completely abrogated the H-2D^d- and Ly49A-dependent inhibition of killing (Matsumoto et al. 2001). The abrogation was a consequence of the inhibition of Ly49A-H-2D^d interaction by the antibody, since the antibody also inhibited the physical interaction between H-2D^d and Ly49A that was artificially expressed on Chinese hamster ovary cells. These data not only suggest the possible involvement of β2m in the Ly49A-H-2D^d interaction, but also suggest that Ly49A might only recognize H-2D^d complexed with mouse β2m but not H-2D^d complexed with β2m from other species including bovine. Because substantial amount of β2m in H-2D^d could be replaced with bovine β2m that is present in FCS used as a medium supplement (Bernabeu et al. 1984). To test the possibility, we have designed an experiment in which a part of the β2m subunit in the cell surface H-2D^d is replaced with human β2m and anti-mouse β2m antibody is included in the functional killing assay (Fig. 1). The result that H-2D^d/human β2m was unable to protect the target cells from killing clearly demonstrates that Ly49A is sensitive to species-specific determinant of β2m and suggest the direct involvement of β2m in the Ly49A-H-2D^d interaction along with α1/α2 domain (Matsumoto et al. 2001). The subsequent demonstration of β2m species-specificity in Ly49A recognition of H-2D^d by introduction of human or mouse β2m into β2m-defective Daudi cells together with H-2D^d heavy chain confirmed the result.

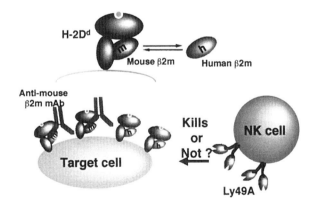

Fig. 1. An experimental system to test the species specific role of β2m in Ly49A recognition of H-2D^d. β2m subunit of H-2D^d is replaced by exogeneously added human β2m. Unreplaced H-2D^d/mouse β2m is masked by anti-mouse β2m antibody to examine the Ly49A recognition of H-2D^d/human β2m.

The functional Ly49A-binding site spans three structural domains of MHC class I

The finding that β2m and β-sheet region of α1/α2 domain are critically involved in the Ly49A recognition of H-2D^d suggests that Ly49A recognition surface might lie beneath α1/α2 domain. However, this idea appeared to be in conflict with the data that an anti-α3 antibody that recognizes a loop in α3 including Lys243 does not affect the recognition. In addition, the two antibodies (34-5-8S and S19.8), which inhibit the recognition bind distinct surfaces of H-2D^d (N. Matsumoto, W.M. Yokoyama, S. Kojima, and K. Yamamoto, submitted for publication; Margulies et al. 1983).

To find the footprint of Ly49A on H-2D^d that can explain these puzzling data, we started to work on a large panel of H-2D^d mutants that have individual Ala substitution on selected solvent-exposed residues. During the course of study, Tormo et al. resolved the crystal structure of the Ly49A/H-2D^d complex (Tormo et al. 1999). The structure provided the two possible binding sites on H-2D^d for Ly49A, designated site 1 and site 2 (Fig. 2A). Site 1 includes NH$_2$-terminal end of α1 α-helix and COOH-terminal end of α2 α-helix of H-2D^d, while site 2 spans the three structural domains of H-2D^d. The above enigmatic data all pointed site 2 as a functional binding site for Ly49A; even though site 1 was suggested to be the functional Ly49A binding site from the geometry of the binding site and polymorphism that exists in site 1 region in the crystal structure paper.

We examined the panel of H-2D^d mutants for Ly49A binding and capacity to protect target cell killing by Ly49A$^+$ NK cells. The panel includes mutants in the residues in site 1 and site 2 of H-2D^d. As our data predicted, Ly49A recognition of H-2D^d was inhibited by individual mutation of four specific residues in site 2 (R6, R111, D122, K243) (Matsumoto et al. 2001). By contrast, single mutations as well as double mutations in site 1 residues did not significantly affect the recognition in both of the binding and functional assays. These data clearly demonstrate that site 2 constitutes the functional binding site for Ly49A that is associated with inhibition of cytotoxicity (Fig. 2A). The data also imply that site 2 is the major binding site for Ly49A.

Fig. 2. The functional binding sites for two structurally distinct NK-cell receptors for MHC class I. (A) H-2D^d mutations in the structure of the Ly49A./H-2D^d complex (PDB id: 1QOX) . H-2D^d is shown in surface model. Ly49A dimers are shown in ribbon model. Each chain of the Ly49A dimers is differently colored. β2m, which contributes species specificity of the Ly49A/H-2D^d interaction, is in pink. Critical residues identified by mutation analysis are red or orange and labeled, Residues that did not affect the interaction after Ala-substitution are green. The images in A, C were prepared with Swiss-PdbViewer (Guex and Peitsch 1997). (B) 16 model structures of N-glycan from CD2 (PDB id: 1GYA) were transplanted to Asn 86 of H-2D^d in a complex with Ly49A. H-2D^d and Ly49A are shown in ribbon model. SYBYL (Tripos Software, St. Louis MO) was used for modeling and preparation of the graphics image. (C) Docking orientation of Ly49A (left, PDB accession code: 1QOX) and KIR2DL2 (right, PDB accession code: 1EFX) onto their MHC class I ligands. The left panel is orthogonal to the right panel with regard to MHC class I.

A

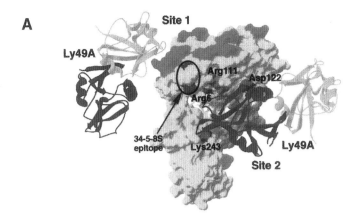

B

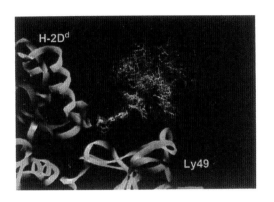

C

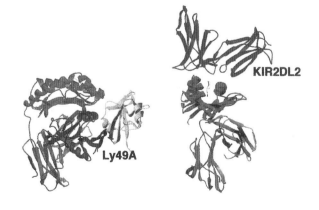

Potential role of carbohydrate in Ly49A recognition of MHC class I

Mouse MHC class I molecules have conserved N-glycosylation sites at Asn 86 and Asn 176. Its receptor Ly49A has an extracellular domain homologous to C-type lectin and indeed has a capacity to bind certain carbohydrate polymers like fucoidan (Daniels et al. 1994). These attempted us to produce and test the glycosylation-deficient H-2D^d for recognition by Ly49A. The results that Ly49A functionally as well as physically bind carbohydrate-deficient H-2D^d have excluded the essential role of carbohydrate moiety of MHC class I in the recognition by Ly49A (Matsumoto et al. 1998). However, several studies suggest that carbohydrate on MHC class I might play some role in the recognition by Ly49A. Therefore, we modeled the complex of Ly49A and H-2D^d glycosylated at Asn 86, which is proximal to the functional binding site for Ly49A. We attempted to transplant a dynamic structure of high mannose-type carbohydrate moiety of CD2 that was resolved by nuclear magnetic resonance spectroscopy (Wyss et al. 1995). We replaced the side chain of H-2D^d Asn 86 with the side chain of CD2 Asn 65, which accompanies N-glycan, reserving conformation of Asn side chain (chi1, chi2) and of N-glycan. In the model N-glycan in any conformation that was found in CD2 was accommodated in the Ly49A/H-2D^d complex without structural optimization process (Fig. 2B). Primarily, the model provides an insight into the stoichiometry of Ly49A binding to H-2D^d. In the Ly49A/H-2D^d complex one Ly49A dimer interacts with two H-2D^d molecules (Tormo et al. 1999). However, the N-glycan modeled on Asn86 of H-2D^d prevents the simultaneous binding of two H-2D^d molecules. Therefore, the model predicts the stoichiometry of the interaction to be one Ly49A dimer to one H-2D^d molecule.

The model with N-glycan also raises the possibility that the N-glycan might interact with the surface of Ly49A. The possible interaction of the N-glycan with Ly49A is consistent with the recently published observation that carbohydrate sulfation might modify the Ly49A recognition of MHC class I (Chang and Kane 1998) and results from the initial studies that used glycosidase and glycosylation inhibitors (Brennan et al. 1995; Daniels et al. 1994).

Interestingly, the site of possible interaction between Ly49A and H-2D^d N-glycan corresponds to carbohydrate recognition surface found in typical carbohydrate binding C-type lectins (Weis et al. 1998). This also implies the following hypothesis to explain the origin of MHC class I recognition by the C-type lectin-like NK cell receptors: Recognition of MHC class I by the precursor of C-type lectin-like NK receptor was originally mediated by interaction between the carbohydrate recognition domain (CRD) of the precursor and the carbohydrates such as those expressed on MHC class I. Then the CRD evolved to acquire the ability to bind protein surface of the MHC class I concomitant with an evolution of C-type lectin-like NK cell receptor in losing its capacity to recognize the carbohydrate residues on MHC class I.

Polymorphic determinants of MHC class I distinguished by Ly49A

Ly49A distinguishes polymorphic MHC class I with a hierarchy (Hanke et al. 1999; Matsumoto, Tajima, Mitsuki, and Yamamoto, submitted for publication). However, the critical residues identified as above are conserved among mouse MHC class I regardless of their capacity to interact with Ly49A. Previous studies including those from our laboratories localized polymorphic determinant in the β-sheet region of α1/α2 domain by use of D^d/K^d recombinant MHC class I molecules (Matsumoto et al. 1998) and in α2 domain by domain swapping between D^d and D^b (Sundback et al. 1998). Mutation studies based on these data have revealed that part of polymorphic determinants lies in peptide-binding groove (Matsumoto, Yokoyama, Kojima, and Yamamoto, submitted for publication; Nakamura et al. 2000; Waldenstrom et al. 1998). Our most recent data suggest the presence of another determinant in the interface of three structural domains that constitute MHC class I: α1/α2 and α3 domains and β2m (Matsumoto, Yokoyama, S. Kojima, Yamamoto, submitted for publication). The data imply that the polymorphic determinant is dependent on configuration of the three structural domains, consistent with the functional Ly49A binding to site 2. However, currently available data are unable to explain all the Ly49A specificity toward polymorphic MHC class I. Further studies are necessary to fully understand the basis for Ly49 discrimination of polymorphism.

KIR recognizes distinct surface of MHC class I from the Ly49A-binding site

Human NK cells express Ig-like receptors KIR to recognize classical MHC class I molecules including HLA-A, B and C (Lanier 1998). The first clue for identifying the KIR binding site was provided by the discrimination of dimorphism of residues in HLA-C (77 and 80) by the two closely related receptors KIR2DL2 and KIR2DL1 (Biassoni et al. 1995). The details of the KIR recognition of HLA-C have recently been unveiled by the crystal structure of the KIR2DL2/HLA-Cw3 complex (Boyington et al. 2000). The KIR binding site lies on top of the MHC class I where the T-cell receptor (TCR) binds to an overlapping but distinct site (Fig. 2C). Despite the differences in structure and in the binding sites on MHC class I, binding of the functionally similar NK-cell receptors is critically mediated by hydrogen bonds and charge-charge interaction as shown by the mutation studies (Boyington et al. 2000; Matsumoto et al. 2001). The overlap of the KIR binding site with that of the TCR precludes the simultaneous recognition of a single MHC class I molecule by the KIR and the TCR; while it is possible in the case of Ly49A.This might have some functional consequences in T cells.

Currently available data suggest different modes of detecting the polymorphic determinants between KIR2DL2 and Ly49A. KIR2DL2 has a direct contact with the polymorphic determinant located at the COOH-terminus of the α1 α-helix (Boyington et al. 2000), while the surface recognized by Ly49A has limited polymorphism. Also, Ly49A rather detects conformational epitope sensitive to substitution in the peptide-binding groove and in the interface of the three structural domains as discussed above. Another difference can be found in the sensitivity to peptide. The KIR recognition of MHC class I is sensitive to the sequence of peptide bound to the groove (Peruzzi et al. 1996) and direct contact of the peptide with KIR has been revealed by the crystal structure. By contrast, Ly49A apparently is insensitive to the sequence of the peptide (Correa and Raulet 1995; Orihuela et al. 1996) and there is no direct contact between Ly49A and peptide in the groove (Tormo et al. 1999). However, peptide-selective recognition of MHC class I ligands has been observed in the other members of Ly49 family (Ly49C, Ly49I) (Franksson et al. 1999; Hanke et al. 1999) and in the other C-type lectin-like receptors CD94/NKG2 (Kraft et al. 2000; Llano et al. 1998). Even though whether these C-type lectin-like receptors recognize a similar region of their MHC class I ligands remains to be determined, the peptide selectivity in these receptors might be related to the presence of the polymorphic determinants in the peptide-binding groove observed for Ly49A.

Despite these differences between the two structurally distinct receptors, both appear to have similar functions to monitor classical MHC class I in their respective species human and mouse. The functional consequences of these differences among the two groups of receptors are unknown. How the two species ended up using structurally distinct receptors is a very interesting question, which remains to be answered.

Acknowledgments

This work was supported by Grants-in-Aid for Scientific Research from the Ministry of Education, Science, and Culture of Japan, by a grant for Research on Health sciences focusing on Drug Innovation from the Japan Health Science Foundation, and by grants from the National Institute of Health to W.M. Yokoyama, who is an Investigator for the Howard Hughes Medical Institute. We gratefully thank R.K. Ribaudo and M.J. Shields for their helpful discussions and reagents, H. Iijima for molecular modeling and discussions, and D.H. Margulies for the coordinates for the Ly49A/H-2D^d complex prior to the release from PDB and discussions.

References

Bernabeu C, van de Rijn M, Lerch PG, Terhorst CP (1984) Nature 308:642-645

Biassoni R, Falco M, Cambiaggi A, Costa P, Verdiani S, Pende D, Conte R, Di Donato C, Parham P, Moretta L (1995) J Exp Med 182:605-609

Bjorkman PJ, Saper MA, Samraoui B, Bennett WS, Strominger JL, Wiley DC (1987) Nature 329:506-512

Boyington JC, Motyka SA, Schuck P, Brooks AG, Sun PD (2000) Nature 405:537-543

Brennan J, Takei F, Wong S, Mager DL (1995) J Biol Chem 270:9691-9694

Chang CS, Kane KP (1998) J Immunol 160:4367-4374

Correa I, Raulet DH (1995) Immunity 2:61-71

Daniels BF, Nakamura MC, Rosen SD, Yokoyama WM, Seaman WE (1994) Immunity 1:785-792

Franksson L, Sundback J, Achour A, Bernlind J, Glas R, Karre K (1999) Eur J Immunol 29:2748-2758

Guex N, Peitsch MC (1997) Electrophoresis 18:2714-2723

Hanke T, Takizawa H, McMahon CW, Busch DH, Pamer EG, Miller JD, Altman JD, Liu Y, Cado D, Lemonnier F A, Bjorkman PJ, Raulet DH (1999) Immunity 11:67-77

Karlhofer FM, Hunziker R, Reichlin A, Margulies DH, Yokoyama WM (1994) J Immunol 153:2407-2416

Karlhofer FM, Ribaudo RK, Yokoyama WM (1992) Nature 358: 66-70

Kraft JR, Vance RE, Pohl J, Martin AM, Raulet DH, Jensen PE (2000) J Exp Med 192:613-624

Lanier LL (1998) Annu Rev Immunol 16:359-393

Llano M, Lee N, Navarro F, Garcia P, Albar JP, Geraghty DE, Lopez-Botet M (1998) Eur J Immunol 28:2854-2863

Margulies DH, Parnes JR, Johnson NA, Seidman JG (1983) Proc Natl Acad Sci USA 80:2328-2331

Matsumoto N, Mitsuki M, Tajima K, Yokoyama WM, Yamamoto K (2001) J Exp Med 143: 147-157

Matsumoto N, Ribaudo RK, Abastado JP, Margulies DH, Yokoyama WM (1998) Immunity 8:245-254

Nakamura MC, Hayashi S, Niemi EC, Ryan JC, Seaman WE (2000) J Exp Med 192:447-454

Orihuela M, Margulies DH, Yokoyama WM (1996) Proc Natl Acad Sci USA 93:11792-11797

Peruzzi M, Parker KC, Long EO, Malnati MS (1996) J Immunol 157:3350-3356

Smith HR, Karlhofer FM, Yokoyama WM (1994) J Immunol 153:1068-1079

Sundback J, Nakamura MC, Waldenstrom M, Niemi EC, Seaman WE, Ryan JC, Karre K (1998) J Immunol 160:5971-5978

Tormo J, Natarajan K, Margulies DH, Mariuzza RA (1999) Nature 402:623-631

Waldenstrom M, Sundback J, Olsson-Alheim MY, Achour A, Karre K (1998) Eur J Immunol 28:2872-2881

Weis WI, Taylor ME, Drickamer K (1998) Immunol Rev 163:19-34

Wong S, Freeman JD, Kelleher C, Mager D, Takei F (1991) J Immunol 147:1417-1423

Wyss DF, Choi JS, Li J, Knoppers MH, Willis KJ, Arulanandam AR, Smolyar A, Reinherz EL, Wagner G (1995) Science 269:1273-1278

Yokoyama WM (1999) Natural killer cells. In: Paul WE (Ed) Fundamental Immunology, Fourth Edition. Lippincott-Raven Publishers, Philadelphia, pp 575-603

28

Characterization of Tm1 cells, a NKR[+] subset of memory-phenotype CD8[+] T cells

Nicolas Anfossi[1], Véronique Pascal[1,2], Sophie Ugolini[4], and Eric Vivier[1,2,3]

[1]Centre d'Immunologie de Marseille-Luminy, CNRS-INSERM-Univ.Med., Campus de Luminy, Case 906, 13288 Marseille cedex 09, France
[2]Service d'Hématologie, Hôpital de la Conception, 13385 Marseille Cedex 05, France
[3]Institut Universitaire de France
[4]Institut de Pharmacologie Moléculaire et Cellulaire, CNRS, 06560 Sophia-Antipolis, France

Summary. ITIM-bearing NK receptors for MHC class I molecules (NKR) can impair NK cell activation programs and confer to NK cells the capacity to discriminate between MHC class I[+] and MHC class I[-] target cells. Inhibitory NKR are thus involved in the control of NK tolerance to self. A subset of T cells also express inhibitory NKR at the cell surface. We refer to these cells as Tm1 cells (T memory type 1 cells), and describe here the phenotypic and functional features of this subset of memory-phenotype CD8[+] T cells. This analysis reveals a novel biological function for inhibitory NKR when expressed on T cells. Indeed, sensing of self-MHC class I molecules by inhibitory NKR displayed on T cell surface leads to the *in vivo* accumulation of Tm1 cells.

Key words. Lymphocyte activation, KIR, LIR, IL-2Rβ, CCR7

Introduction

αβT cell maturation and activation of effector function is dependent upon engagement of the T cell antigen receptor complex (TCR) by peptide-loaded self-MHC molecules. MHC class I molecules can also interact with inhibitory Natural Killer cell Receptors (NKR) (Lanier 1998; Long 1999; Moretta et al. 2000; Tomasello et al. 2000). Via inhibitory NKR, NK cells can sense the alteration of MHC class I molecules at the surface of target cells. In the absence of appropriate interaction between inhibitory NKR and MHC class I molecules, NK cells can exert their cytolytic function. Therefore, NK cells eliminate autologous cells that

225

present quantitative and/or qualitative alterations of MHC class I molecules, such as virus-infected cells as well as tumor cells. Inhibitory NKR include receptors for classical MHC class Ia molecules, such as Killer cell Ig-like Receptors (KIR) in humans and lectin-like Ly49 receptors in mice. In humans, the Leukocyte Ig-like Receptor-1 (LIR-1) also serves as a low affinity inhibitory receptor for multiple MHC class I molecules, via the recognition of the relatively non-polymorphic MHC class I α3 domain (Chapman et al. 1999). Finally, in human and mouse, the CD94/NKG2A heterodimers are lectin-like inhibitory receptors that serve as receptors for the non classical MHC class Ib molecules (HLA-E in human and Qa-1b in mice). On NK cells, the inhibitory function of NKR involves intracytoplasmic Immunoreceptor Tyrosine-based Inhibition Motifs (ITIM) which recruit the protein tyrosine phosphates SHP-1 and/or SHP-2 (Daëron and Vivier 1999; Long 1999; Ravetch and Lanier 2000)

Despite the well-described in vitro inhibition of T cell clone effector function by NKR engagement, the biological function of NKR on T cells in vivo remains to be elucidated. We will review the features of memory-phenotype $CD8^+$ T cells expressing inhibitory NKR, a subset that we refer to as Tm1 cells thereafter.

Tm1 cells are IL-2Rβ$^+$ memory-phenotype CD8$^+$ T cells

Cell surface phenotype of human Tm1 cells

A discrete subset of peripheral blood αβT cells express KIR and/or LIR-1 (Mingari et al. 1998; Speiser et al. 1999). KIR^+ T cells have been defined as $CD28^-CD45RA^-CD45R0^+$ T cells, a phenotype which is enriched in memory $CD8^+$ T cells (Mingari et al. 1996). Extensive cell surface phenotype of $CD3^+KIR^+$ T cells revealed that these cells are $CD2^+CD5^+CD7^+CD29^+CD49d^+CD122^+2B4^+$ T cells (Table 1). The vast majority of these cells also express the CD8 molecule at their surface. Despite the ~20% of $CD3^+KIR^+$ T cells which are $CD8^-$, only ~7% of $CD3^+KIR^+$ T cells are $CD4^+$, implying that a fraction of $CD3^+KIR^+$ T cells are $CD4^-CD8^-$. $CD4^-CD8^-$ T cells expressing NK markers, NK/T cells, are characterized by the cell surface expression of semi-invariant TCR (see below) and have been extensively characterized (Chiu et al. 1999). Therefore, we will focus our analysis on $CD8^+KIR^+$ T cells.

Several conclusions can be drawn from the cell surface phenotype analysis of circulating human $CD3^+KIR^+$ T cells, and can be extended to all human Tm1 cells ($CD3^+KIR^+$ T cells and $CD3^+LIR-1^+$ T cells).

First, memory T cell subsets include central memory T cells (T_{CM}: $CCR7^+CD45RA^-$) which migrate to lymphoid organs, as well as effector memory T cells (T_{EM}: $CCR7^-CD45RA^-$) which migrate to inflamed non-lymphoid tissues

(Sallusto et al. 1999). The lack of CCR7 expression on $CD3^+KIR^+$ T cells suggests that Tm1 cells belong the T_{EM} cell subset. T_{EM} cell divide into T cells which migrate to the lungs and to the skin (CLA^+, $\alpha4\beta1^+$, $CCR4^+$) as well as into T cells which migrate into the gut ($\alpha4\beta7^+$, $CCR9^+$). $CD3^+KIR^+$ T cells are homogeneously $\alpha4^+$T cells. Two distinct integrins include the CD49d/$\alpha4$ polypeptide. The $\alpha4\beta1$ integrin (VLA-4) is a receptor for the extracellular matrix protein fibronectin and the endothelial adhesion protein VCAM-1. $\alpha4\beta1$ is involved in the homing of memory and effector T cells to inflamed tissues, and especially the lung. The $\alpha4\beta7$ integrin is another receptor for fibronectin but also serves as a receptor for the mucosal addressin-cell adhesion molecule 1 (MadCAM 1). $\alpha4\beta7$ is involved in the homing of lymphocytes to gut and associated lymphoid tissues. $CD3^+KIR^+$ T cells express VLA-4, consistent with the migratory properties of T_{EM} cells. However, circulating Tm1 cells do not express the $\alpha E\beta7$ integrin (CD103), and are therefore phenotypically distinct from resident intra-epithelial lymphocytes characterized by the cell surface expression of $\alpha E\beta7$.

Second, Tm1 cells uniformly express the 2B4 (CD244) activation receptor as well as the common β chain for IL-2 and IL-15 oligomeric complexes (IL-2Rβ/IL-15Rβ, CD122). 2B4, a ligand for CD48, is not only expressed on all NK cells, but also on a subset of T cells which have NK-like killing properties (Brown et al., 1998). IL-2Rβ is also expressed on all NK cells, and serves as a marker for a subset of memory $CD8^+$ T cells (Tough et al. 1999). Together with the expression of BY55 (CD160) on a fraction of Tm1 cells, these results support the notion that Tm1 cells are memory $CD8^+$ T cells capable of immediate cytotoxic function. Indeed, the expression of the BY55 protein has been shown to be tightly associated with NK and $CD8^+$ T lymphocytes with cytolytic effector activity (Anumanthan et al. 1998).

Third, heterogeneity in the expression of CD56, CD57 and CD27 molecule suggest that Tm1 cells might include cells at various stages of activation, but further experiments are clearly needed to investigate this issue.

Fourth, direct evidence that Tm1 cells are antigen-experienced T cells was obtained using HLA class I tetramers (HLA-A2) loaded with antigenic peptides derived from Epstein Barr virus (EBV: BMLF1 epitope) and human Cytomegalovirus (CMV: pp65 epitope). A large proportion of peptide-loaded HLA-A2 tetramer$^+$ T cells expresses LIR-1 at their surface (up to 80%), as compared to the total $CD8^+$ T cell population (data not shown).

Finally, the frequency of Tm1 cells increases with age, as a result of increased frequencies of both $CD8^+KIR^+$ T cells and $CD8^+LIR-1^+$ T cells, but not of $CD8^+CD94/NKG2A^+$ T cells (Ugolini et al. submitted). These results contrast with previous data showing that the number of KIR^+ T cells was constant over time, and confirm that Ig-like NKR (KIR and LIR-1) and lectin-like NKR (CD94/NKG2A) are regulated by distinct mechanisms (André et al. 1999). Although Tm1 cells are barely detectable in young infants, the size of the Tm1 subset can reach up to 50% of total CD8+ T cells in elderly individuals.

Altogether, these data characterize human Tm1 cells as a population of antigen-experienced CD8$^+$ T cells harboring an IL-2Rβ^+ T$_{EM}$ phenotype.

Table 1. Phenotype of human Tm1 cells

Cell surface molecule	% CD3$^+$KIR$^+$ cells	
	Mean ± SEM	Range
CD2	100 ± 0	100 - 100
CD29/β1 integrin	100 ± 0	100 - 100
CD49d/α4 integrin	100 ± 0	100 - 100
CD122/IL-2Rβ	100 ± 0	100 - 100
CD244/2B4	100 ± 0	100 - 100
CD5	94 ± 1	91 - 96
CD7	93 ± 2	89 - 99
CD8	77 ± 6	67 - 93
CD57	59 ± 9	25 - 88
CD62L	48 ± 8	31 - 70
CD56	44 ± 17	27 - 78
CD160/BY55	30 ± 9	11 - 51
CD27	25 ± 9	5 - 76
CD26	15 ± 1	12 - 18
CD28	9 ± 2	0.4 - 20
CD4	7 ± 3	3 - 16
CD38	6 ± 3	3 - 12
CD16	1.4 ± 1.4	0 - 4
CD23	0 ± 0	0- 0
CD25	0 ± 0	0- 0
CD103/αEβ7	0 ± 0	0 - 0
CDw197/CCR7	0 ± 0	0 - 0
CLA	0 ± 0	0 - 0

The cell surface phenotype of Tm1 cells was assessed by three-color flow cytometry gating on CD3$^+$KIR$^+$ T cells. Results are expressed as the mean % ± SEM of CD3$^+$KIR$^+$ T cells expressing the indicated cell surface molecules as well as the range of variations observed among peripheral blood samples collected from 4 to 7 healthy volunteers.

Cell surface phenotype of mouse Tm1 cells

A subset of mouse T cells in peripheral blood, in spleen, in lymph nodes and in the liver express Ly49 molecules and harbor a CD8$^+$ memory phenotype (Coles et al. 2000; Norris et al. 1999). In contrast to NK cells, several groups have reported that T cells only express the inhibitory Ly49 molecules (Coles et al. 2000; Ortaldo et al. 1998). However, it has been recently described that T cells may also express activating Ly49 molecules (e.g. Ly49D) at their surface (Kambayashi et al. 2000). In humans, activating KIR molecules can be expressed on T cell clones (Cambiaggi et al. 1996; Vély et al. in press). Irrespective of this issue, mouse Tm1 cells share several important features with human Tm1 cells. Indeed, they harbor an IL-2Rβ$^+$CD25$^-$CD69$^-$ phenotype and express the markers of mouse memory-phenotype CD8$^+$ T cells: CD44^{high} and Ly6C (Coles et al. 2000). In addition, the size of the Tm1 subset increases with age, and reaches up to 20% of total CD8$^+$ T cells in aged mice.

Induction of NKR expression on T cells

In both humans and mice, Tm1 cells are not detected in thymus. Tm1 cells are also not detectable in human cord blood nor in mouse fetal liver (Mingari et al. 1998). Similar to NK cells, the mechanisms which govern the induction of inhibitory NKR expression at the T cell surface remain to be elucidated. However, the isolation of several panels of human T cell clones expressing identical TCR, but various KIR phenotypes (such as KIR$^+$ vs. KIR$^-$) formally proves that the induction of inhibitory NKR expression occurs after termination of *TCRA/TCRB* genes recombination events (Vély et al. in press). It has been also reported that TCR engagement in vitro might up-regulate KIR expression on human T cell clones (Huard and Karlsson 2000). Together with the memory-phenotype of Tm1 cells, these data suggest that the induction of inhibitory NKR on the T cell surface occurs upon T cell activation. The nature of this activation and whether it is dependent upon antigenic stimulation is under investigation. In this regard, IL-21 (in combination with flt-3L and IL-15) is the only cytokine described so far as capable of inducing high surface levels of CD16 on CD56$^+$ NK cells derived from human CD34 bone marrow progenitors (Parrish-Novak et al., 2000). As KIR are only expressed on mature-phenotype CD56^{dim}CD16^{bright} NK cells (André et al. 2000), it will be of interest to investigate whether IL-21 is involved in the induction of KIR on NK and T cells.

NKR/MHC class I-dependent survival of Tm1 cells

While inhibitory NKR have been involved in NK cell self-tolerance, the biological function of inhibitory NKR on T cells remains to be precisely elucidated. In vitro experiments performed on T cell clones have suggested that NKR might act as inhibitors of T cell effector function (D'Andrea et al. 1996; Ikeda et al. 1997; Mingari et al. 1998; Mingari et al. 1997; Mingari et al. 1996; Mingari et al. 1995; Phillips et al. 1995; Speiser et al. 1999). However, our previous in vivo results have shown that the role of NKR on T cells might be more complex than originally thought. Indeed, using transgenic mice (Tg-KIR/HLA) expressing a single inhibitory human NKR (KIR2DL3) and its cognate HLA class I ligand (HLA-Cw3), we could not detect any alteration of thymic development, T cell repertoire (CDRβ3 size distribution) and in vivo primary T cell response as compared to control mice (wild type, Tg-KIR and Tg-HLA)(Cambiaggi et al. 1999). Despite the lack of inhibition of T cell responses in Tg-KIR/HLA mice, NK cells from Tg-KIR as well as from Tg-KIR/HLA mice are tolerant to H-2 mismatch targets expressing HLA-Cw3, indicating that the transgenic KIR2DL3 receptors can function as inhibitory molecules in mouse lymphocytes in vitro and in vivo (Cambiaggi et al. 1997). We have recently analyzed the memory CD8[+] T cell compartment in Tg-KIR/HLA mice. Our results show that engagement of inhibitory NKR selectively drives the in vivo accumulation of Tm1 cells (Ugolini et al. submitted). In addition, our in vitro experiments suggest that inhibition of activation-induced cell death is a potential mechanism to explain the MHC class I-dependent in vivo accumulation of NKR[+] T cells (Ugolini et al. submitted).

These results describe an unexpected biological function for inhibitory NKR on T cells, as well as a novel strategy of MHC class I-dependent memory-phenotype CD8[+] T cell survival in the periphery. As NKR are much less sensitive to peptide identity than the TCR, and interact with the products of multiple MHC class I alleles (Long 1999), our results are consistent with a previous report showing that the in vivo survival of HY-specific memory CD8[+] T cells requires the peptide-independent recognition of a variety of MHC class I molecules (Tanchot et al. 1997). However, it has been recently demonstrated that LCMV-specific memory CD8[+] T cells can persist in the absence of MHC class I (Murali-Krishna et al. 1999). Heterogeneity of memory-phenotype CD8[+] T cells has been previously described (Oehen and Brduscha-Riem 1998; Sallusto et al. 1999). It is thus possible that LCMV-specific and HY-specific memory CD8[+] T cells which survive or disappear in MHC class I[+] or in MHC class I[-] environments belong to distinct memory CD8[+] T cell subsets. Interestingly, it has been recently reported that a subset of antigen-reactive Tm1 cells expand upon influenza A virus infection (Kambayashi et al. 2000). It thus remains to determine whether and which parameters of the antigenic stimulation (quality, quantity, site of exposure) might dictate the initiation of a Tm1 response, as well as whether Tm1 cells represent a separate lineage or a maturation step of memory CD8[+] T cells.

The expression pattern of CCR7 indicates that Tm1 cells belong to the T_{EM} subset, suggesting that sustained and/or repetitive antigen exposure leads to the development of a Tm1 response. Along this line, it has been shown that the cell surface expression of IL-2Rβ on antigen-experienced CD8[+] T cells correlates with chronic antigenic exposure (Bieganowska et al. 1999). The high frequency of Tm1 cells within the pool of CD8$^+$ T cells chronically exposed to antigenic stimulation, e.g. CD8$^+$LIR-1$^+$IL-2Rβ^+ in EBV- or CMV-healthy carriers (data not shown), further supports that chronic T cell stimulation might preferentially lead to the development of a Tm1 response. In this state of chronic stimulation, the fine tuning of TCR signals by NKR could prevent clonal exhaustion of antigen-stimulated T cells by raising their activation threshold. As NKR-induced inhibitory signals can be bypassed by strong activation signals (Bléry et al. 1997), Tm1 cell effector function is likely to be operative during an acute antigenic challenge.

Concluding remarks

We describe here the feature of Tm1 cells as well as the involvement of ITIM-bearing MHC class I receptors in the survival of this subset of memory-phenotype CD8$^+$ T cells. Other T cell subsets have been described to express cell surface receptors which had been initially characterized on NK cells. In particular, NK1.1$^+$ (NKRP-1C) NKT cells express semi-invariant TCR (i.e. Vα14 in the mouse and Vα24 in humans), and recognize glycolipids such as α-galactosylceramide presented by CD1d (Chiu et al. 1999). NKT cells account for 15% of mature mouse thymocytes and <1% mouse splenocytes. NKT are also characterized by a CD4$^-$CD8$^-$ or a CD4$^+$CD8$^-$ cell surface phenotype. Tm1 cells markedly differ from NKT cells by their phenotype, their distribution and the absence of a restricted TCR repertoire. Tm1 cells include CD8$^+$KIR$^+$ and CD8$^+$LIR-1$^+$ T cells in humans as well as CD8$^+$Ly49$^+$ T cells in the mouse. It remains to be defined whether CD8$^+$CD94/NKG2A$^+$ T cells belong to the Tm1 cell subset. Indeed, no age-dependent increase in CD94/NKG2A$^+$CD8$^+$ T cells could be detected (Ugolini et al. submitted), and distinct mechanisms regulate the cell surface expression of KIRs/LIR-1 and CD94/NKG2A (André et al. 1999).

References

André P, Brunet C, Guia S, Gallais H, Sampol J, Vivier E, Dignat-George F (1999) Differential regulation of killer cell Ig-like receptors and CD94 lectin-like dimers on NK and T lymphocytes from HIV-1-infected individuals. Eur J Immunol 29:1076-1085

André P, Spertini O, Guia S, Rihet P, Dignat-George F, Brailly H, Sampol J, Anderson PJ, Vivier E (2000) Modification of P-selectin glycoprotein ligand-1 with a natural killer cell-restricted sulfated lactosamine creates an alternate ligand for L-selectin. Proc Natl Acad Sci USA 97:3400-3405

Anumanthan A, Bensussan A, Boumsell L, Christ AD, Blumberg RS, Voss SD, Patel AT, Robertson MJ, Nadler LM, Freeman GJ (1998) Cloning of BY55, a novel Ig superfamily member expressed on NK cells, CTL, and intestinal intraepithelial lymphocytes. J Immunol 161:2780-2790

Bieganowska K, Hollsberg P, Buckle GJ, Lim DG, Greten TF, Schneck J, Altman JD, Jacobson S, Ledis SL, Hanchard B, Chin J, Morgan O, Roth PA, Hafler DA (1999) Direct analysis of viral-specific CD8+ T cells with soluble HLA-A2/Tax11-19 tetramer complexes in patients with human T cell lymphotropic virus-associated myelopathy. J Immunol 162:1765-1771

Bléry M, Delon J, Trautmann A, Cambiaggi A, Olcese L, Biassoni R, Moretta L, Chavrier P, Moretta A, Daëron M, Vivier E (1997) Reconstituted killer-cell inhibitory receptors for MHC Class I molecules control mast cell activation induced via immunoreceptor tyrosine-based activation motifs. J Biol Chem 272:8989-8996

Brown MH, Boles K, van der Merwe AP, Kumar V, Mathew PA, Barclay NA (1998) 2B4, the natural killer and T cell immunoglobulin superfamily surface protein, is a ligand for CD48. J Exp Med 188:2083-2090

Cambiaggi A, Darche S, Guia S, Kourilsky P, Abastado J P, Vivier E (1999) Modulation of T-cell functions in KIR2DL3 (CD158b) transgenic mice. Blood 94:2396-2402

Cambiaggi A, Orengo A M, Meazza R, Sforzini S, Tazzari P L, Lauria F, Raspadori D, Zambello R, Semenzato G, Moretta L, Ferrini S (1996) The natural killer-related receptor for HLA-C expressed on T cells from CD3+ lymphoproliferative disease of granular lymphocytes displays either inhibitory or stimulatory functions. Blood 87:2369-2375

Cambiaggi A, Verthuy C, Naquet P, Romagne F, Ferrier P, Biassoni R, Moretta A, Moretta L, Vivier E (1997) Natural killer cell acceptance of H-2 mismatch bone marrow grafts in transgenic mice expressing HLA-Cw3 specific killer cell inhibitory receptor. Proc Natl Acad Sci USA 94:8088-8092

Chapman TL, Heikeman AP, Bjorkman PJ (1999) The inhibitory receptor LIR-1 uses a common binding interaction to recognize class I MHC molecules and the viral homolog UL18. Immunity 11:603-613

Chiu YH, Jayawardena J, Weiss A, Lee D, Park SH, Dautry-Varsat A, Bendelac A (1999) Distinct Subsets of CD1d-restricted T Cells Recognize Self-antigens Loaded in Different Cellular Compartments. J Exp Med 189:103-110

Coles MC, McMahon CW, Takizawa H, Raulet DH (2000) Memory CD8 T lymphocytes express inhibitory MHC-specific Ly49 receptors. Eur J Immunol 30:236-244

D'Andrea A, Chang C, Phillips JH, Lanier LL (1996) Regulation of T cell lymphokine production by killer cell inhibitory receptor recognition of self HLA class I alleles. J Exp Med 184:789-794

Daëron M, Vivier E (1999) Biology of immunoreceptor tyrosine-based inhibition motif bearing molecules. Curr Top Microbiol Immunol 244:1-12

Huard B, Karlsson L (2000) KIR expression on self-reactive CD8+ T cells is controlled by T-cell receptor engagement. Nature 403:325-328

Ikeda H, Lethé B, Lehmann F, Van Baren N, Baurain J-F, De Smet C, Chambost H, Vitale M, Moretta A, Boon T, Coulie PG (1997) Characterization of an antigen that is recognized on a melanoma showing partial HLA loss by CTL expressing an NK inhibitory receptor. Immunity 6:199-208

Kambayashi T, Assarsson E, Michaëlsson J, Berglund P, Diehl AD, Chambers BJ, Ljunggren H-G (2000) Emergence of CD8+ T cells expressing NK cell receptors in influenza A virus-infected mice. J Immunol 165:4964-4969

Lanier L L (1998) NK Cell Receptors. Annu Rev Immunol 16:359-393

Long E O (1999) Regulation of immune responses through inhibitory receptors. Annu Rev Immunol 17:875-904

Mingari MC, Moretta A, Moretta L (1998) Regulation of KIR expression in human T cells: a safety mechanism that may impair protective T-cell responses. Immunol Today 19:153-157

Mingari MC, Ponte M, Cantoni C, Vitale C, Schiavetti F, Bertone S, Bellomo R, Tradori Cappai A, Biassoni R (1997) HLA-class I-specific inhibitory receptors in human cytolytic T lymphocytes: molecular characterization, distribution in lymphoid tissues and co-expression by individual T cells. Int Immunol 9:485-491

Mingari MC, Schiavetti F, Ponte M, Vitale C, Maggi E, Romagnani S, Demarest J, Pantaleo G, Fauci AS, Moretta L (1996) Human CD8+ T lymphocyte subsets that express HLA class I-specific inhibitory receptors represent oligoclonally or monoclonally expanded cell populations. Proc Natl Acad Sci USA 93:12433-12438

Mingari MC, Vitale C, Cambiaggi A, Schiavetti F, Melioli G, Ferrini S, Poggi A (1995) Cytolytic T lymphocytes displaying natural killer (NK)-like activity: expression of NK-related functional receptors for HLA class I molecules (p58 and CD94) and inhibitory effect on the TCR-mediated target cell lysis or lymphokine production. Int Immunol 7:697-703

Moretta L, Biassoni R, Bottino C, Mingari MC, Moretta A (2000) Human NK-cell receptors. Immunol Today 21:420

Murali-Krishna K, Lau LL, Sambhara S, Lemonnier F, Altman J, Ahmed R (1999) Persistence of Memory CD8 T Cells in MHC Class I-Deficient Mice. Science 286:1377-1381

Norris S, Doherty DG, Collins C, McEntee G, Traynor O, Hegarty JE, O'Farrelly C (1999) Natural T cells in the human liver: cytotoxic lymphocytes with dual T cell and natural killer cell phenotype and function are phenotypically heterogenous and include Valpha24-JalphaQ and gammadelta T cell receptor bearing cells. Hum Immunol 60:20-31

Oehen S, Brduscha-Riem K (1998) Differentiation of naive CTL to effector and memory CTL: correlation of effector function with phenotype and cell division. J Immunol 161:5338-5346

Ortaldo J, Winkler-Pickett R, Mason AT, Mason LH (1998) The Ly-49 family: regulation of cytotoxicity and cytokine production in murine $CD3^+$ cells. J Immunol 160:1158-1165

Parrish-Novak J et al. (2000) Interleukin 21 and its receptor are involved in NK cell expansion and regulation of lymphocyte function. Nature 408:57-63

Phillips JH, Gumperz JE, Parham P, Lanier LL (1995) Superantigen-dependent, cell-mediated cytotoxicity inhibited by MHC class I receptors on T lymphocytes. Science 268:403-405

Ravetch JV, Lanier LL (2000) Immune inhibitory receptors. Science 290:84-89

Sallusto F, Lenig D, Forster R, Lipp M, Lanzavecchia A (1999) Two subsets of memory T lymphocytes with distinct homing potentials and effector functions. Nature 401:708-712

Speiser DE, Pittet MJ, Valmori D, Dunbar R, Rimoldi D, Lienard D, MacDonald HR, Cerottini JC, Cerundolo V, Romero P (1999) In vivo expression of natural killer cell inhibitory receptors by human melanoma-specific cytolytic T lymphocytes. J Exp Med 190:775-782

Tanchot C, Lemonnier FA, Pérarnau B, Freitas AA, Rocha B (1997) Differential requirements for survival and proliferation of CD8 naïve or memory T cells. Science 276:2057

Tomasello E, Bléry M, Vély E, Vivier E (2000) Signaling pathways engaged by NK cell receptors: double concerto for activating receptors, inhibitory receptors and NK cells. Semin Immunol 12:139-147

Tough DF, Sun S, Zhang X, Sprent J (1999) Stimulation of naive and memory T cells by cytokines. Immunol Rev 170:39-47

Ugolini S, Arpin C, Anfossi N, Walzer T, Cambiaggi A, Forster R, Lipp M, Toes RE, Melief CJ, Marvel J, Vivier E (submitted) Involvement of inhibitory NKR in the survival of Tm1 cells, a NKR+ subset of memory-phenotype CD8+ T cells.

Vély F, Peyrat M-A, Couedel C, Morcet J-F, Halary F, Davodeau F, Romagne F, Scotet E, Saulquin X, Houssaint E, Schleinitz N, Moretta A, Vivier E, Bonneville M (2001) Regulation of inhibitory and activating Killer-cell-Ig-like receptor (KIR) expression occurs in T cells after termination of TCR rearrangement. J Immunol in press

29
How do killer cell Ig-like receptors inhibit natural killer cells?

Eric O. Long[1], Deborah N. Burshtyn[2], Christopher C. Stebbins[1], and Carsten Watzl[1]

[1]Laboratory of Immunogenetics, National Institute of Allergy and Infectious Diseases, National Institutes of Health, 12441 Parklawn Drive, Rockville, Maryland 20852, USA
[2]Department of Medical Microbiology and Immunology, Heritage Medical Research Centre, University of Alberta, Edmonton, Alberta T6G 2S2, Canada

Summary. Several receptors expressed on natural killer (NK) cells provide negative regulation of the NK cell's cytotoxic activity. These inhibitory receptors carry immunoreceptor tyrosine-based inhibition motifs (ITIM) in their cytoplasmic tail that are responsible for the negative signal. ITIM-mediated inhibition of cellular signals has become a common theme in many different cell types that express a receptor with one or more cytoplasmic ITIM. Despite much work on the MHC class I-specific inhibitory receptors expressed by NK cells, the mechanism by which activation signals are blocked is still not well understood.

We have chosen to study the inhibition of NK cells by the killer cell Ig-like receptor (KIR) during contact with target cells that express a class I ligand for KIR. Engagement of KIR by HLA-C on the target cell resulted in complete inhibition of the tyrosine phosphorylation of the activation receptor 2B4. In addition, the formation of stable cell-cell conjugates between NK cells and target cells was aborted by KIR binding to HLA-C on the target cells. Therefore, KIR inhibits activation of NK cells at a step as proximal as the phosphorylation of an activation receptor and the adhesion to target cells.

Key words. Natural killer, Inhibitory receptor, Adhesion, Tyrosine phosphorylation

Introduction

The cytotoxic activity of NK cells is completely blocked by KIR if the target cell expresses an MHC class I ligand for KIR. Expression of MHC class I molecules on hematopoietic cells is necessary to protect them from lysis by NK cells. Other cell types exhibit various degrees of sensitivity to NK cells and may rely less, or not at all on MHC class I for protection from lysis by NK cells. The reason may be that the many activation receptors expressed by NK cells recognize tissue-specific ligands.

KIR inhibit NK cells by recruitment of the tyrosine phosphatase SHP-1 (Burshtyn et al. 1996). Phosphorylation of the tyrosine in the two V/IxYxxL ITIM sequences results in specific binding to the tandem SH2 domains of SHP-1, thereby activating the SHP-1 catalytic activity (Burshtyn et al. 1997). The ITIM sequence has since appeared in a large number of receptors, many of which have inhibitory functions (Ravetch and Lanier 2000). At least superficially, the inhibitory property of KIR was explained by the ITIM-mediated recruitment of SHP-1. Activation of NK cytotoxicity is dependent on tyrosine kinases of the src family and on the syk tyrosine kinase. Dephosphorylation of tyrosine residues by activated SHP-1 through coligation or colocalization of KIR with activation receptors would abort the signaling cascade initiated by these tyrosine kinases.

Important questions about the mechanism of inhibition by KIR remain. How are the ITIMs phosphorylated? Although KIR blocks tyrosine kinase-dependent pathways, it must itself serve as substrate of a tyrosine kinase before it can exert its inhibition. Whether KIR depends on phosphorylation in *trans* by a kinase associated with activation receptors has not been established. If so, it would explain why inhibitory receptors act only locally at the site of coligation with activation receptors (Bléry et al. 1997). Is SHP-1 selective about its substrate targets or does it dephosphorylate several components of the activation pathway? Experiments using antibodies to crosslink ITIM-containing receptors with activation receptors revealed a global reduction in tyrosine phosphorylated proteins, when compared to crosslinking of the activation receptor (e.g. CD16) alone (Binstadt et al. 1996, 1998; Palmieri et al. 1999). This finding is compatible with either a selective dephosphorylation of a key proximal component in the signaling pathway or a nonselective dephosphorylation of multiple substrates.

During cell-cell contact, the formation of a tight conjugate induces important changes in cell shape, in the distribution of surface molecules, and in the organization of membrane microdomains, all of which impact greatly on the signals that are transduced. Therefore, a true appreciation of the KIR inhibitory function can only be derived from analysis of direct NK cell-target cell interactions. Our approach has been to study the effect of KIR in NK cells that are mixed with target cells expressing or not a ligand for KIR. We wish to eventually determine the precise requirements for tyrosine phosphorylation of the ITIM and for the inhibitory signal. We have so far focused on two aspects of the inhibition exerted by KIR: its effect on the tyrosine phosphorylation of an activation receptor, and on the formation of NK-target cell conjugates.

Several earlier studies of signal inhibition by KIR during ligation by MHC class I on target cells established some important points (Kaufman et al. 1995; Valiante et al. 1996). First, the step at which NK activation is blocked is upstream of calcium flux. Second, the inhibition must be exerted at the level of selected signals because the overall level of protein tyrosine phosphorylation in the NK cells was not greatly reduced during inhibition by KIR. A few unknown proteins had a reduced level of tyrosine phosphorylation, and the linker for activation of T cells (LAT) did not associate with Grb2, suggesting that it was not phosphorylated (Valiante et al. 1996). In a γ/δ T cell clone expressing the ITIM-containing receptor CD94/NKG2A the tyrosine phosphorylation of ZAP70 and of an unknown p130 protein, but not that of the T cell receptor ζ chain, was reduced if the antigen-presenting cell expressed an MHC class I ligand for the inhibitory receptor (Carena et al. 1997). These results contrast the marked reduction in overall tyrosine phosphorylation induced by crosslinking KIR and activation receptors with antibodies. Whether KIR and other ITIM-containing receptors target unique, selected substrates, or multiple substrates for dephosphorylation by SHP-1, and whether SHP-1 substrates lie at a proximal or a distal step in the signaling pathway, are still open questions.

Results

KIR engagement by HLA-C on target cells prevents the tyrosine phosphorylation of the activation receptor 2B4 (CD244)

2B4 is a member of the CD2 subfamily of the Ig superfamily and is expressed by all NK cells and by some T cells. It acts as an activation receptor for NK cells (Valiante and Trinchieri 1993) and its signal can synergize with signals from ITAM-containing receptors (Sivori et al. 2000). The cytoplasmic tail of 2B4 contains four tyrosine residues within the sequence context TxYxxI/V. This sequence motif is similar to that of three tyrosine residues in the cytoplasmic of the T cell activation molecule CD150 (SLAM). As shown for CD150, 2B4 can associate with the single SH2 domain-containing molecule SAP (also called SH2D1A) or with the tyrosine phosphatase SHP-2, but not both at the same time (Tangye et al. 1999). NK cells from patients with mutations in SAP, who suffer from X-linked lymphoproliferative (XLP) disease, are not activated by 2B4, implying a role of SAP in the 2B4 activation signal (Nakajima et al. 2000; Parolini et al. 2000; Tangye et al. 2000; Benoit et al. 2000). However, the mechanism by which 2B4 and CD150 transmit a positive signal is still unknown. The ligand of 2B4 is CD48, a glycophospholipid-anchored surface protein expressed on most hematopoietic cells. It was of interest to test the inhibitory effect of KIR on 2B4 because inhibition of non-ITAM activation signals had not been reported.

Crosslinking 2B4 on NK cells with mAb C1.7 (Valiante and Trinchieri 1993) resulted in tyrosine phosphorylation of 2B4 and of several other proteins. Therefore, phosphorylation of 2B4 does not require ligation of other activation receptors. Mixing of NK cells with the cell line 721.221 that expresses CD48 also resulted in rapid tyrosine phosphorylation of 2B4. The NK cell line YTS transfected with the HLA-Cw4-specific KIR2DL1 (YTS-2DL1) was used to test how KIR engagement would affect 2B4 phosphorylation. As expected, mixing of YTS-2DL1 with 221-Cw3 cells that are not protected from killing still resulted in rapid tyrosine phosphorylation of 2B4. In contrast, mixing with 221-Cw4 cells that are not killed did not result in 2B4 phosphorylation. This was not simply due to a lack of interaction between YTS-2DL1 and 221-Cw4 cells because tyrosine phosphorylation of SHP-1 was detected after cell mixing. Inhibition of 2B4 phosphorylation occurs also in normal NK cells. A population of IL-2-activated NK cells isolated from a donor that expresses an inhibitory form of CD94 (CD94/NKG2A) on the majority of its NK cells was used to test the effect of CD94 ligation on 2B4. To test how ligation of the ITIM-containing CD94/NKG2A by HLA-E would affect 2B4 phosphorylation we mixed these NK cells with the target cells 221-Cw3 and 221-Cw7. The HLA-Cw3 and –Cw7 allotypes are recognized by the same KIR but cell surface HLA-E is expressed only in the 221-Cw3 cells. 2B4 phosphorylation was greatly reduced when NK cells were mixed with the 221-Cw3 cells, showing that inhibition of 2B4 phosphorylation occurs also in normal NK cells. These data have been published in Watzl et al. (2000).

In conclusion, the tyrosine phosphorylation of 2B4 induced by incubation with target cells expressing CD48 is prevented by the binding of inhibitory receptors to MHC class I on the target cells. This could occur by dephosphorylation of 2B4 by SHP-1 or, alternatively, by the dephosphorylation of a SHP-1 substrate that lies even upstream of 2B4 phosphorylation in the NK activation pathway. In either case, the data show that NK inhibitory receptors can interfere with a step as proximal as tyrosine phosphorylation of an activation receptor.

KIR engagement inhibits formation of NK cell conjugates with target cells

A prerequisite for target cell killing by cytotoxic T and NK cells is the formation of a tight cell-cell conjugate. The β2 integrin LFA-1 (CD11a/CD18) plays a major role in adhesion of T and NK cells to target cells. An 'inside-out' signal to LFA-1 is necessary for the formation of a high affinity interaction of LFA-1 with its ligand ICAM-1 on target cells. The requirement of signal transduction in the process of adhesion to target cells represents a potential step at which KIR may exert its inhibition. Blocking adhesion would be an effective way to prevent NK activation at an early step.

A two-color flow cytometry assay was used to quantitate the formation of tight conjugates between the cell line YTS and the target cell line 721.221. At a YTS:221 ratio of 1:2, 80% of YTS cells formed conjugates with 221 cells within less than 10 minutes. Identical results were obtained with YTS cells mixed with 221-Cw3 or 221-Cw4 cells, and YTS-2DL1 cells mixed with 221-Cw3 cells. In contrast, adhesion to 221-Cw4 cells was interrupted abruptly within less than 5 minutes by engagement of the inhibitory receptor on YTS-2DL1. Masking KIR2DL1 with a specific IgM mAb prevented the inhibitory effect, thereby implicating this receptor in the inhibition of adhesion. Is inhibition of adhesion mediated by recruitment of SHP-1? To address this point we used a chimeric receptor consisting of KIR2DL1 fused to an intact SHP-1. SHP-1 was linked to the cytoplasmic tail of KIR2DL1 upstream of the first ITIM and the chimeric construct was expressed in the cell line YTS. As control, a catalytically inactive SHP-1, unable to cleave or bind substrates, was also linked to KIR2DL1. The KIR2DL1/SHP-1 chimera was very effective at blocking conjugate formation with 221-Cw4 cells, whereas the chimera with the mutated SHP-1 was not. Therefore, SHP-1 is sufficient to mediate inhibition of adhesion.

A number of inhibitors and antibodies were used to evaluate the requirements for the formation of NK:target cell conjugates. A blocking anti-CD11a antibody inhibited YTS:221 conjugate formation. A blocking anti-CD28 antibody and soluble CTLA-4 (a high affinity ligand of B-7) each inhibited adhesion only partially under conditions that blocked lysis completely. Therefore, the interaction of CD28, expressed by YTS, with B-7 on 221 cells is more important for target cell lysis than it is for conjugate formation. Inhibition of tyrosine kinase by PP1 or by herbimycin A resulted in fewer conjugates, and in very little target cell lysis. Conjugate formation was temperature dependent, showing a noticeable decrease at 22°C. Taken together, these data indicate that signal transduction is necessary for tight adhesion between YTS and 221 cells. These data have been published in Burshtyn et al. (2000).

Although the pathway for inside-out signaling to LFA-1 is not known, it is clear that it offers a point at which KIR can interfere with the cytotoxic activity of NK cells. This function of KIR could serve two purposes: inhibition of lysis, and release of the NK cell from a resistant target cell that should not be killed. The latter may be useful to reduce the time spent by NK cells contacting normal cells that have to be spared.

Conclusions

Two models can be proposed. In the first, recruitment and activation of SHP-1 by KIR results in dephosphorylation of tyrosine residues on several components of the overall activation pathway that leads to NK cytotoxicity. For example, SHP-1 may dephosphorylate the activation receptor 2B4, the tyrosine kinase syk, and other effectors such as PLCγ or proteins that induce the high affinity form of LFA-1. Such a model would explain how KIR can block receptor phosphorylation, calcium signals, and adhesion to target cells. However, the model has to account for earlier findings that engagement of KIR by MHC class I on target cells results in a selective reduction of protein tyrosine phosphorylation. While there could be multiple substrate targets of SHP-1 during inhibition by KIR, global dephosphorylation of tyrosine residues is not consistent with available data. The crystal structure of the SHP-1 catalytic domain complexed with phosphotyrosine peptides and biochemical measurements on rates of peptide dephosphorylation showed that the amino acid sequence context of the phosphotyrosine has a major influence on catalytic activity (Yang et al., 2000). Dephosphorylation of a few selected substrates by SHP-1, possibly in sequential order, is still a viable model.

According to the second model, activated SHP-1 recruited by KIR exerts its inhibition of NK activation by dephosphorylating a single phosphotyrosine protein. That protein would evidently be essential in signaling pathways leading to adhesion and cytotoxicity of NK cells. The observed inhibition of 2B4 phosphorylation and of adhesion to target cells would be the manifestation of that protein's central role.

A distinction between the two models could only be obtained by identification of the direct substrate(s) of SHP-1 during NK inhibition by KIR. A definitive answer will be provided by analysis of SHP-1 substrates in intact NK cells right when they are experiencing the potent veto power of KIR. Elucidation of the inhibitory mechanism used by KIR to block NK activation could help understand the function of those many other ITIM-containing receptors that regulate a wide range of cellular responses.

References

Benoit L, Wang X, Pabst HF, Dutz J, Tan R (2000) Defective NK cell activation in x-linked lymphoproliferative disease. J Immunol 165: 3549-3553

Binstadt, BA, Brumbaugh KM, Dick CJ, Scharenberg AM, Williams BL, Colonna M, Lanier LL, Kinet JP, Abraham RT, Leibson PJ (1996) Sequential involvement of Lck and SHP-1 with MHC-recognizing receptors on NK cells inhibits FcR-initiated tyrosine kinase activation. Immunity 5:629-638

Binstadt BA, Billadeau DD, Jevremovic D, Williams BL, Fang N, Yi TL, Koretzky GA, Abraham RT, Leibson PJ (1998) SLP-76 is a direct substrate of SHP-1 recruited to killer cell inhibitory receptors. J Biol Chem 273:27518-27523

Bléry M, Delon J, Trautmann A, Cambiaggi A, Olcese L, Biassoni R, Moretta L, Chavrier P, Moretta A, Daëron M, Vivier E (1997) Reconstituted killer cell inhibitory receptors for major histocompatibility complex class I molecules control mast cell activation induced via immunoreceptor tyrosine-based activation motifs. J Biol Chem 272:8989-8996

Burshtyn DN, Scharenberg AM, Wagtmann N, Rajagopalan S, Berrada K, Yi T, Kinet JP, Long EO (1996) Recruitment of tyrosine phosphatase HCP by the killer cell inhibitor receptor. Immunity 4:77-85

Burshtyn DN, Yang WT, Yi TL, Long EO (1997) A novel phosphotyrosine motif with a critical amino acid at position-2 for the SH2 domain-mediated activation of the tyrosine phosphatase SHP-1. J Biol Chem 272:13066-13072

Burshtyn DN, Shin J, Stebbins C, Long EO (2000) Adhesion to target cells is disrupted by the killer cell inhibitory receptor. Curr Biol 10:777-780

Carena I, Shamshiev A, Donda A, Colonna M, Libero GD (1997) Major histocompatibility complex class I molecules modulate activation threshold and early signaling of T cell antigen receptor-gamma/delta stimulated by nonpeptidic ligands. J Exp Med 186:1769-1774

Kaufman DS, Schoon RA, Robertson MJ, Leibson PJ (1995) Inhibition of selective signaling events in natural killer cells recognizing major histocompatibility complex class I. Proc Natl Acad Sci USA 92:6484-6488

Nakajima H, Cella M, Bouchon A, Grierson HL, Lewis J, Duckett CS, Cohen JI, Colonna M (2000) Patients with X-linked lymphoproliferative disease have a defect in 2B4 receptor-mediated NK cell cytotoxicity. Eur J Immunol 30:3309-3318

Palmieri G, Tullio V, Zingoni A, Piccoli M, Frati L, Lopez-Botet M, Santoni A (1999) CD94/NKG2-A inhibitory complex blocks CD16-triggered Syk and extracellular regulated kinase activation, leading to cytotoxic function of human NK cells. J Immunol 162:7181-7188

Parolini S, Bottino C, Falco M, Augugliaro R, Giliani S, Franceschini R, Ochs HD, Wolf H, Bonnefoy JY, Biassoni R, Moretta L, Notarangelo LD, Moretta A (2000) X-linked lymphoproliferative disease: 2B4 molecules displaying inhibitory rather than activating function are responsible for the inability of natural killer cells to kill Epstein-Barr virus-infected cells. J Exp Med 192:337-346

Ravetch JV, Lanier LL (2000) Immune inhibitory receptors. Science 290:84-89

Sivori S, Parolini S, Falco M, Marcenaro E, Biassoni R, Bottino C, Moretta L, Moretta A (2000) 2B4 functions as a co-receptor in human NK cell activation. Eur J Immunol 30:787-793

Tangye SG, Lazetic S, Woollatt E, Sutherland GR, Lanier LL, Phillips JH (1999) Cutting edge: Human 2B4, an activating NK cell receptor, recruits the protein tyrosine phosphatase SHP-2 and the adaptor signaling protein SAP. J Immunol 162:6981-6985

Tangye SG, Phillips JH, Lanier LL, Nichols KE (2000) Functional requirement for SAP in 2B4-mediated activation of human natural killer cells as revealed by the X-linked lymphoproliferative syndrome. J Immunol 165: 2932–2936

Valiante NM, Trinchieri G (1993) Identification of a novel signal transduction surface molecule on human cytotoxic lymphocytes. J Exp Med 178:1397-1406

Valiante NM, Phillips JH, Lanier LL, Parham P (1996) Killer cell inhibitory receptor recognition of human leukocyte antigen (HLA) class I blocks formation of a pp36/PLC-gamma signaling complex in human natural killer (NK) cells. J Exp Med 184:2243-2250

Yang J, Cheng ZL, Niu TQ, Liang XS, Zhao ZZJ, Zhou GW (2000) Structural basis for substrate specificity of protein-tyrosine phosphatase SHP-1. J Biol Chem 275:4066-4071

Watzl C, Stebbins CC, Long EO (2000) Cutting edge: NK cell inhibitory receptors prevent tyrosine phosphorylation of the activation receptor 2B4 (CD244). J Immunol 165:3545-3548

Index

DATE DUE